h care and increasing agricultural products to ıcing a cleaner environment. The public markets nise this potential, and a lot of money has flowed ›iotechnology companies.

ıis book discusses the cutting edge developments in eld of biotechnology and its' impacts in society. It es readers through multi-staged biotechnical cations, and explores biotechnology's effects on 's society and on future generations. Besides students, ers and researchers of biotechnology, the contents of ›ok will also be valuable to all who wish to improve knowledge in biotechnology.

Editor

MODERN TRENDS IN BIOTECHNOLOGY

Dr. Hema Gautam

RAJAT PUBLICATIONS
New Delhi

Modern Trends in Biotechnology
by *Dr. Hema Gautam*

This edition is published by Rajat Publications
4740/23, Ansari Road Daryaganj,
New Delhi - 110002

ISBN: 978-81-7880-265-7

Edition 2025

Printed and bound in India

RAJAT
PUBLICATIONS

4740/23, Ansari Road, Daryaganj,
New Delhi- 110002 (India)
Mob:- +91-9811484973, +91-9811105389
E-mail: rajat_publication@yahoo.com

Preface

Biotechnology seems to be leading a su
revolution. It has brought us to the b
"engineered" products that are based i
rather than on chemical and ind
Biotechnology has been described as
implies that there are two sides. On or
DNA to be manipulated to move genes
to another. On the other, it involv
technologies whose consequences are
be met with caution.

Biotechnology is currently being
including agriculture, bioremediation,
energy production. DNA fingerprin
common practice in forensics. Produ
other medicines is accomplished throu
that now carry the chosen gene. Imm
not only in medicine for drug level ar
but also by farmers to aid in detectio
pesticides, herbicides and toxins on
products. These assays also provide
industrial chemicals in ground water
In agriculture, genetic engineering is b
plants that are resistant to insects
diseases. Biotechnology-derived
potential to positively change our soc

heal
proc
reco
into

T
the
guid
appl
toda
teach
the b
their

MODERN TRENDS IN BIOTECHNOLOGY

Dr. Hema Gautam

RAJAT PUBLICATIONS
New Delhi

Modern Trends in Biotechnology
by *Dr. Hema Gautam*

This edition is published by Rajat Publications
4740/23, Ansari Road Daryaganj,
New Delhi - 110002

ISBN: 978-81-7880-265-7

Edition 2025

Printed and bound in India

RAJAT
PUBLICATIONS

4740/23, Ansari Road, Daryaganj,
New Delhi- 110002 (India)
Mob:- +91-9811484973, +91-9811105389
E-mail: rajat_publication@yahoo.com

Preface

Biotechnology seems to be leading a sudden new biological revolution. It has brought us to the brink of a world of "engineered" products that are based in the natural world rather than on chemical and industrial processes. Biotechnology has been described as "Janus-faced". This implies that there are two sides. On one, techniques allow DNA to be manipulated to move genes from one organism to another. On the other, it involves relatively new technologies whose consequences are untested and should be met with caution.

Biotechnology is currently being used in many areas including agriculture, bioremediation, food processing, and energy production. DNA fingerprinting is becoming a common practice in forensics. Production of insulin and other medicines is accomplished through cloning of vectors that now carry the chosen gene. Immunoassays are used not only in medicine for drug level and pregnancy testing, but also by farmers to aid in detection of unsafe levels of pesticides, herbicides and toxins on crops and in animal products. These assays also provide rapid field tests for industrial chemicals in ground water, sediment, and soil. In agriculture, genetic engineering is being used to produce plants that are resistant to insects, weeds and plant diseases. Biotechnology-derived products have the potential to positively change our society, from improving health care and increasing agricultural products to producing a cleaner environment. The public markets recognise this potential, and a lot of money has flowed into biotechnology companies.

This book discusses the cutting edge developments in the field of biotechnology and its' impacts in society. It guides readers through multi-staged biotechnical applications, and explores biotechnology's effects on today's society and on future generations. Besides students, teachers and researchers of biotechnology, the contents of the book will also be valuable to all who wish to improve their knowledge in biotechnology.

Editor

Contents

1 Modern Agriculture

Science is an elegant way of getting at the truth, according to science writer Rick Weiss. It should follow then that molecular biology and other tools of modern biotechnology add elegance and precision to the pursuit of solutions to thwart poverty, malnutrition and food insecurity in too many countries around the world. In agriculture these enemies are manifest as pests, diseases, drought and other biotic and abiotic stresses that limit the productivity of plants and animals. But not all appreciate the elegance of science in the pursuit of truth.

Mendel's laws of genetics were rediscovered in 1900. Mendel had published his work on inheritance patterns in pea in 1865, but it took 35 years for others to grasp their significance. Since 1900, steady progress in our understanding of the genetic makeup of all living organisms ranging from microbes to humans. A major step in human control over genetic traits was taken in the 1920s when Muller and Stadler discovered that radiation can induce mutations in animals and plants.

In the 1930s and 1940s, several new methods of chromosome and gene manipulation were discovered, such as the use of colchicine to achieve a doubling in chromosome number, commercial exploitation of hybrid vigor in maize and other crops, use of chemicals such as nitrogen mustard and ethyl methane sulphonate to induce mutations, and techniques such as tissue culture and embryo rescue to make viable hybrids from distantly related species.

The double helix structure of DNA, the chemical substance of heredity, was discovered in 1953 by James Watson and Francis Crick. This triggered explosive progress in every field of genetics. In 21st century, there is a rapid transition from Mendelian to molecular genetic applications in agriculture, medicine, and industry. This brief capsule of genetic progress from 1900 to 1999 stresses that knowledge and discovery represent a continuum, with each generation taking understanding of the complex web of life to a higher level. It would therefore be a mistake to worship or discard experimental tools or scientific innovations because they are either old or new. Just as it took 35 years for biologists to understand fully the significance of Mendel's work, it may take a couple of decades more to understand fully the benefits and risks associated with new genetically improved organisms.

The 1990s have seen dramatic advances in understanding of how biological organisms function at the molecular level, as well as in our ability to analyze, understand, and manipulate DNA molecules, the biological material from which the genes in all higher organisms are made. The entire process has been accelerated by the Human Genome Project, which has invested substantial public and private resources into the development of new technologies to work with human genes. The same technologies are directly applicable to other organisms, including plants and animals. Thus, the new scientific discipline of genomics has arisen, which has contributed to powerful new approaches to identify the functions of genes and their application in agriculture and medicine.

These new discoveries and their commercial application have helped to promote the biotechnology industry, mainly in North America and Europe.

Several large corporations in Europe and the United States have made major investments to adapt these technologies to produce improved plant varieties of agricultural importance for large-scale commercial agriculture. The same technologies have equally important potential applications to address food security and poverty of people in developing countries. The current use of modern biotechnology in Asia, Latin America and the Caribbean, sub- Saharan Africa, and West Asia/North Africa was reviewed at an international conference in October 1999 in Washington. The current status of agricultural biotechnology in China, India, the Philippines, and Thailand was reviewed at the conference. All are committed to the use of modern biotechnology in agriculture, and are investing significant human and financial resources to this policy and have done so over the past decade.

China sees the greatest challenge as the use of biotechnology to increase food production and improve product quality in an environmentally sustainable manner. China has moved quickly to adopt new technologies. Over 103 genes have been evaluated for improving traits in 47 plant species. The crops include rice, wheat, corn, cotton, tomato, pepper, potato, cucumber, papaya, and tobacco. A variety of traits were targeted for improvement including disease resistance, insect resistance, herbicide resistance, and quality improvement. Approximately 50 genetically improved organisms (GIOs) have been approved for commercial production, environmental release or small-scale field testing in China. In a few cases, new genetically improved varieties have been approved for large-scale commercial production. These are being grown commercially on approximately 1 million hectares of land in China in 1999. It is expected that the area planted will increase rapidly in the next few years.

India has allocated large public resources toward infrastructure and human resources development in biotechnology. Current efforts are toward applications in improving agricultural productivity; bioremediation in the environment; medical biotechnology for the production of new vaccines, diagnostics and drugs; industrial biotechnology; and bioinformatics. Research and development (R&D) priorities in agriculture include new regeneration protocols for rapid multiplication of citrus, coffee, mangrove, vanilla and cardamom. Yield of cardamom has increased 40 percent using tissue-cultured plants.

Thailand is focusing on the applications of biotechnology to traditional foods, fruits and export commodities. R&D priorities are to raise production and cut costs by using new biotechnology on crops such as rice, sugarcane, rubber, durian, and orchids. An early success in Thailand has been in the application of biotechnology to develop new molecular diagnostics for the diagnosis and control of virus diseases in shrimps. These diseases cost the shrimp export industry over US$500 million in lost production in 1996.

The Philippines began its biotechnology programs in 1980 with the creation of the National Institutes of Molecular Biology and Biotechnology, with a focus on agricultural biotechnology. In 1997, the Agriculture Fisheries Modernization Act recognized biotechnology as a major strategy to increase agricultural productivity. The Act will provide a budget for agricultural biotechnology of almost US$20 million annually for the next 7 years, an increase from US$1 million per year. In 1998, five high level biotechnology research projects were funded by government: Development of new varieties of banana resistant to banana bunchy top virus and papaya resistant to ringspot virus; delayed ripening of papaya and mango; insectresistant corn; marker-assisted breeding in coconut; and coconut with high lauric acid content. Public concerns have been vocal in the Philippines and this is constraining the commercial use of modern biotechnology in agriculture.

All four countries have regulatory systems in place at the national and institutional level to govern R&D programs and commercial developments where appropriate. Intellectual property management was considered to be a difficult issue for all four countries. Agricultural biotechnology was reviewed in Brazil, Costa Rica, and Mexico, as examples of the diversity of uses and views of biotechnology across the region. The main challenges identified for the region were: management of intellectual property in relation to major and minor crops; assessment of several research options, not only a molecular approach, in assessing how best to tackle problems and challenges to improve agricultural productivity; identification of beneficiaries; prioritization of work on favored and/or marginal areas; use of GIOs as indicators of environmental damage; and need to monitor the behavior of GIOs in the environment after release.

In Brazil, many lines of research and development are benefiting from the application of biotechnology tools such as marker-assisted plant and animal breeding, genomic mapping of several species including sugarcane, embryo transfer applied to different animal species, genetic resources characterization and conservation, and use of genetic improvement to introduce new traits, such as papaya resistant to papaya ring spot virus and beans resistant to golden mosaic virus. The issues of field testing of genetically improved plants need to be addressed. Tropical agriculture is very different from the temperate fields where most of the new genetically improved products have been tested. Protocols are required for field trials, risk assessment, registration of products, and public acceptance. The need is urgent, because these are constraints that will intensify as new genetically improved organisms become an integral part of the research agenda in the region.

Mexico was one of the first developing countries to begin the evaluation of genetically improved plants in the field, commencing in 1988, with trials of plants genetically improved for insect resistance, virus resistance, and delayed ripening. Some of these materials such as the virus-resistant potato are now being grown by Mexican farmers. There is a current debate in Mexico as to the desirability of testing genetically improved corn in Mexico, as it is the center of origin of the crop and wild relatives occur. Regulatory officials in Mexico are interested in continuing field trials with such new varieties if there is a clear benefit to Mexican agriculture, such as new varieties of corn with tolerance to aluminum. This phenotype has great potential to reclaim for agriculture tropical acid soils that were lost due to high levels of soluble aluminum ions. Future field evaluations of corn are likely to be accompanied by a scientific monitoring program after release.

In Costa Rica, there is particular interest in using the tools of biotechnology to characterize and conserve biodiversity. Costa Rican institutions have developed some innovative partnerships for bioprospecting, which could serve as a model for other countries. In agriculture, pesticide use has increased three fold between 1993 and 1996, on crops such as banana, coffee, and rice. Much of this is used to control banana diseases where the excess pesticide use is leading to poisoning of field workers and contamination of land, water, and animals. Biotechnology-based solutions are urgently needed to replace chemical control of banana diseases. New virus resistance is being introduced into local rice varieties.

The current status of policy and programs using biotechnology in Kenya, South Africa, and Zimbabwe were reviewed and were complemented by interventions on the situation in other African countries. The major challenge identified was the persistent poor performance of agriculture in Africa, which is leading to a food and a poverty crisis. The issues concerning many countries are how to improve food security, increase productivity, conserve biodiversity, reduce pest management costs, deal with increasing urban migration, and reduce poverty. Specific issues related to biotechnology are how to develop institutional capacity for risk assessment and management, to access information on developments in biotechnology elsewhere that may have application in Africa, and to develop the necessary human resources and infrastructure.

Several success stories are coming out of Africa, where biotechnological approaches have contributed to the solution of specific problems, reduced the cost of pest control, and created new employment opportunities in towns and villages. They include the wide adoption by farmers of rapid multiplication of disease-free banana plantlets in Kenya; use of new genetically improved pest-resistant cotton varieties by small farmers in South Africa; and use of new vaccines against animal diseases in Kenya and Zimbabwe.

Some of the problems and constraints identified include: lack of awareness of the benefits and risks associated with modern biotechnology; lack of capacity in some countries to deal with assessing these benefits and risks and in regulating the use of modern biotechnology; high investment costs associated with biotechnological innovations; and increasing concerns being expressed in the media about the potential negative impacts of biotechnology and the need for public awareness of the issues. In sub-Saharan Africa the need is both to improve awareness and institutional capacity to develop biotechnology-based products and, perhaps as importantly, for African scientists and policymakers to articulate an African scientific agenda and to participate in critical global debates on trade, economic growth, and the role of science.

One of the major targets for biotechnology in Egypt is the production of transgenic plants conferring resistance to biotic and abiotic stresses, which are major agricultural problems leading to serious yield losses in many economically important crops in Egypt. Agricultural Genetic Engineering Research Institute, (AGERI) was established in 1990 with the aim of mobilizing the most recent technologies available worldwide to address problems facing agricultural development.The challenges identified were the need to increase agricultural productivity while preserving the fragile natural resource base in the region, and the need to conserve the rich indigenous plant and animal species. The opportunities include: using modern biotechnology to develop crop varieties tolerant to biotic and abiotic stresses, especially drought and salt tolerance; improving the nutritional quality of agricultural commodities; producing biofertilizers and biopesticides; and improving the availability of soil nutrients.

The main constraints in the region are inadequate financial resources, lack of qualified personnel, poor infrastructure, and insufficient regional and international collaboration. There is also a lack of clear strategies, policies, and regulatory frameworks to guide the use of modern biotechnology in most countries of the region. The outcomes of all the regional reviews summarized above stressed the importance of making safe and effective use of the new developments in modern biotechnology to solve agricultural problems. The application of biotechnological advances could help countries deal with the major challenges of feeding expanding populations from the existing land and water resources, and also to alleviate poverty by stimulating the growth of bioscience-based industries, with an associated growth in employment.

Developing countries that are pursuing the safe and effective use of modern biotechnology recognized the need to have in place effective regulatory systems at the national and institutional levels, compatible with international best practice. There was also a widely held view that intellectual property management is a major issue, in terms of enabling access to other people's technology and in stimulating rewarding local innovation. Countries need to have sufficient knowledgeable people to deal with these issues nationally, and in international negotiations and bilateral contractual arrangements.

There is considerable work in progress in the use of modern biotechnology in developing countries, probably much more than is commonly recognized internationally. Much of the R&D in developing countries is being funded by national governments, and some by bilateral and multilateral development agencies. It is estimated that at least twice as much of the biotechnology-related R&D in the public sector in developing countries is being funded by national governments, as by the international development community at present. Only a small amount comes from the private sector. The total R&D effort is still substantially less than the high private sector investments in biotechnology in North America, Europe, and other Organization for Economic Cooperation and Development countries. Nevertheless, there is movement toward a critical mass of public sector investments in biotechnology in several developing countries.

The greater specificity in the handling of genes has meant that inventors could protect their discoveries by means of patents and other forms of intellectual property rights. This has led to an explosion of private investment in the biosciences in the last 15 years, leading to a "biotechnology revolution." The greatest number of modern biotechnology applications are in health care, where they offer new hope to patients with AIDS, genetically inherited diseases, diabetes, influenza, and some forms of cancer. Biotechnology-based processes are now used routinely in the production of most new medicines, diagnostic tools, and medical therapies. The global market for these products in 1998 was approximately US$13 billion.

New developments in agricultural biotechnology are being used to increase the productivity of crops, primarily by reducing the costs of production by increasing the efficiency of use and decreasing the need for inputs of pesticides and herbicides, mostly in crops grown in temperate zones. The applications of agricultural biotechnology are developing new strains of plants that give higher yields with fewer inputs, can be grown in a wider range of environments, give better rotations to conserve natural resources, provide more nutritious harvested products that keep much longer in storage and transport, and continued low cost food supplies to consumers. Private industry has dominated research, accounting for approximately 80 percent of all R&D. Consolidation of the industry has proceeded rapidly since 1996, with more than 25 major acquisitions and alliances worth US$15 billion.

During the past decade, the commercial cultivation of transgenic plant varieties has commenced. In 1999, it is estimated that approximately 40 million hectares of land were planted with transgenic varieties of over 20 plant species, the most commercially important of which were cotton, corn, soybean, and rapeseed. The value of the global market in transgenic crops grew from US$75 million in 1995 to US$1.64 billion in 1998. The traits these new varieties contain include insect resistance (cotton, corn), herbicide resistance (corn, soybean), and delayed fruit ripening (tomato). The benefits of these new crops are better weed and insect control, higher productivity, and more flexible crop management. These benefits accrue primarily to farmers and agribusinesses, although there are also economic benefits accruing to consumers in terms of maintaining food production at low prices. Health benefits for consumers are also emerging from new varieties of corn and rapeseed with modified oil content and reduced levels of potentially carcinogenic mycotoxins.

The broader benefits to the environment and the community through reduced use of pesticides contribute to a more sustainable agriculture and better food security. Other crop/input trait combinations presently being field-tested include virus-resistant melon, papaya, potato, squash, tomato, and sweet pepper; insect-resistant rice, soybean, and tomato; disease-resistant potato; and delayed-ripening chili pepper. There also is work in progress to use plants such as corn, potato, and banana as minifactories for the production of vaccines and biodegradable plastics. Further scientific advances will likely result in crops with a wider range of traits, some of which are likely to be of more direct interest to consumers, for example, by having traits that confer improved nutritional quality to food. Crops with improved output traits could confer nutritional benefits to millions of people who suffer from malnutrition and deficiency disorders. Genes have been identified that can modify and enhance the composition of oils, proteins, carbohydrates, and starch in food/feedgrains and root crops. For example, a gene encoding beta carotene/vitamin A formation has been incorporated experimentally in rice. This would enhance the diets of the 180 million children who suffer from the vitamin A deficiency that leads to 2 million deaths annually. Similarly, introducing genes that increase available iron levels in rice three-fold is a potential remedy for iron deficiency that affects more than 2 billion people and causes anemia in about half that number.

Applications of biotechnology in agriculture are in their infancy. Most current genetically improved plant varieties are modified only for a single trait, such as herbicide tolerance or pest resistance. The rapid progress being made in genomics may enhance plant breeding as more functional genes are identified. This may enable more successful breeding for complex traits such as drought and salt tolerance, which are controlled by many genes. This would be of great benefit to those farming in marginal lands worldwide, because breeding for such traits has had limited success with conventional breeding of the major staple food crops.

Genetic Engineering of Crops

Genetic engineering of crops became available in the early 1980's as genetic engineering techniques were being perfected. The advantage of genetic engineering is that it allows the transfer of a single gene, or a couple of genes, in a much more precise, controllable and predictable way than is achievable with conventional breeding. Conventional breeding involves random mixing of the tens of thousands of genes present in a plant with the ten of thousands of genes present in another plant. In contrast, genetic engineering is a much more precise way of improving crops. Improvements can be achieved in much shorter time frames. In the past, it took 10 to 15 years to introduce a new variety of sweet corn. This time has been cut in half using the tools of genetic engineering. In addition, genetically engineered crops are required to undergo extensive food and environmental safety assessment, whereas conventionally modified crops are not.

Today, over 800 million people face daily hunger; furthermore, a majority of the global population growth in the next 50 years will be in developing countries where malnutrition is already prevalent. Forty percent of the world's land use for agriculture is already seriously degraded. In order to meet the nutritional needs of this growing population, cereal production alone will need to increase by 40% in the next 20 years. We simply cannot achieve the kinds of yield increases in a sustainable way using traditional methods of breeding. Biotechnology is an important tool in addition to all of the other tools to produce a food supply that will be sustainable in the long run and will be able to meet these needs in the future.

Almost all of the world's major crops are now being improved using genetic engineering, although initial efforts have focused on commodity crops-primarily corn, soybeans, cotton, canola and potatoes. However, there are many other crops, including other cereal grains and a number of vegetables and fruits, that have been improved using genetic engineering and are in the commercialization pipeline (Table 1).

Biotechnology crops are the most rapidly adopted technology in the history of U.S. agriculture. The first genetically modified crops were commercially grown in 1996. USDA estimates for 2001 indicate that about 25 million acres (20% of global total) of corn, 82 million acres of soybeans (63%), 7 million acres of cotton (13%) and 3 million acres of canola (5%) grown globally are genetically modified varieties. FDA has approved 51 different products that are in various stages of commercialization. One of the first products of plant biotechnology was insect resistant corn. A single gene encoding an insecticidal protein from Bacillus thuringiensis (Bt) was transferred into corn. The Bt gene confers resistance to the European corn borer, a devastating insect pest. In addition to the European corn borer, corn earworm is another problem in sweet corn, and both need to be controlled.

Table 1. Types of Crops Genetically Modified with Quality Traits and Approved for U.S. Field Tests Through 2001

Crop	Quality Characteristic
Alfalfa	Decreased lignin, delayed ripening
Apple	Sugar alcohols increased, altered ripening, carbohydrate metabolism altered
Barley	Heat stable enzymes, novel protein, storage protein altered, disulfide bonds reduced, digestibility improved
Broccoli	Prolonged shelf-life
Canola	Altered fatty acid profile, increased lysine, altered amino acid composition
Carrot	Improved nutrition
Cassava	Improved nutrition
Coffee	Delayed ripening, reduced caffeine
Corn	Altered carbohydrate metabolism, increased methionine, lysine and tryptophan, altered amino acid composition, improved protein quality, altered oil profile, decreased phytate, decreased lignin, increased phosphate, anthocyanins in seeds, mycotoxin production inhibition
Cotton	Improved fiber strength and quality
Grape	Improved fruit quality
Lettuce	Brown spot resistance, delayed senescence
Melon	Delayed ripening
Papaya	Delayed ripening

Pear	Altered ripening
Peppers	Delayed ripening, increased shelf-life
Pineapple	Increased sweetness
Plum	Delayed ripening
Potato	Increased solids, increased carbohydrates, improved nutritional quality, increased tyrosine, altered amino acids, black spot bruise resistance, decreased glycoalkaloids
Rice	Increased starch, altered protein storage, novel protein produced
Soybean	Improved protein quality, increased methionine, lysine, improved animal feed quality, novel protein introduced, altered oil quality, increased phytosterols and phytostanols, fumonisin degradation
Strawberry	Delayed ripening
Sunflower	Altered storage protein, animal feed quality improvement
Sweet potato	Altered amino acid composition, protein quality improved
Tomato	Delayed ripening, increased starch and solids, altered sugar profile, pigments altered, fruit size enlarged, altered carotenoid content, altered processing characteristics, protein quality altered, increased antioxidant enzymes
Wheat	Storage protein altered, improved digestibility, increased methionine, nutritional quality improved, novel protein introduced

Consumers typically throw away ears of sweet corn if they find worms, even though the heavy use of pesticides assures that the worms do not survive more than about an inch down the cob. Consumers who find worms in their frozen or canned corn are sure to register complaints with the company. To avoid these problems, growers, particularly in the southern states that grow two or three crops of corn each year, may spray sweet corn up to 60 times in a season to control these pests. Adoption of this technology would result in a higher quality product and significant reduction in the use of pesticides.

Although the European corn borer and corn earworm have the word "corn" in their names, they affect many other vegetables, such as green beans and peppers. The same Bt gene in field corn can be transferred into other vegetables and confer resistance to those same, or related, insect pests. The Bt gene has been transferred into potatoes to confer resistance to the Colorado potato beetle, a devastating pest in potatoes. The leaves of conventional potato plants are almost totally denuded by the Colorado potato beetle, and this lowers crop yields. The Bt gene also has been transferred into cotton. The cotton boll in genetically modified cotton is protected from the cotton boll weevil; thus, yields are significantly higher. Cotton and potato crops receive extensive pesticide treatment in the United States, and use of the genetically modified varieties results in a reduction in pesticides.

The second major area of genetic modification is herbicide tolerance. Herbicide tolerant crops are tolerant to the more environmentally friendly herbicides, for example, glyphosate. Glyphosate and other broad-spectrum herbicides eliminate a variety of weeds, so the farmer can use one product instead of several different chemicals. These products also significantly reduce crop injury, and, because they degrade quickly in the soil, farmers can be less concerned about crop rotation. One of the most valuable aspects of these products from an environmental perspective is that they encourage "no till" farming methods, which significantly reduce soil erosion.

The vegetable processing industry would value the same herbicide tolerance in sweet peas. Weeds growing in pea fields are harvested along with the peas. Pea fields are frequently contaminated with "nightshade," a weed that has a berry the same size, color and density as peas. When these berries are thermally processed along with the peas, the berries harden and are highly unacceptable to consumers. Farmers have a hard time eliminating nightshade from their pea fields; herbicide tolerant sweet peas would simplify weed control and improve final product quality.

Farmers have recognized the benefits of crop biotechnology and have readily adopted it. Use of these crops has resulted in environmental and food safety benefits, including a significant 2.7 million pounds per year reduction in the use of pesticides, and modeling studies indicate that ground water contamination by herbicides can be significantly reduced when glyphosate tolerant crops are planted. An additional benefit of Bt corn is that, because the worms cannot form holes in the corn, fungi that can produce toxic mycotoxins, e.g., fumonisin, cannot attack the corn, resulting in a cleaner, safer product.

Biotechnology and Rural Poor

There are now a wide range of public, private and NGO organizations with differing objectives attempting to deliver appropriate products to different groups of farmers. This has significant implications regarding the nature of the technology disseminating organizations that plant biotechnologists interact with in identifying what priority technologies are needed, what farmers are the resultant client group and what types of farmers and consumers will ultimately reap the benefits of plant biotechnology research and development.

Agricultural extension is now in a process of reform and transition world-wide. Pressures towards cost-recovery and privatization have led to rapid ,slimming and of public sector extension services in Europe, the USA and Australasia over the last decade. In parallel, public sector agricultural extension services in developing countries are achieving only limited impact but face unsustainably high recurrent costs. In many countries, governments are withdrawing NARs from extension services and now expects other institutions to provide and/or finance such activities. Financial pressures have led to the search for ways of reducing public sector costs by e.g. privatizing parts of the extension service, having farmers pay government for some services, and cost-sharing arrangements between government and NGOs or farmers' organizations.

The most efficient public sector extension services of the future are likely to focus on spheres inadequately serviced by the private commercial sector. As a result novel extension approaches are emerging which are approaches which are participatory, institutionally pluralistic and geared towards cost-sharing. A range of approaches for farmer-led approaches to agricultural extension have been presented. It is unclear how plant biotechnology research could better interface with such changes, especially in relation to NGO or farmer-led approaches to agricultural research and extension. It cannot be assumed that even useful agricultural biotechnologies which are wholly in the public domain will actually reach the fields of poorer farmers in the short term through existing state extension channels.

The Gatsby Charitable Foundation has recognized this in developing a 'research-managed extension' (RME) model for more effective transfer of agricultural biotechnologies to poorer farmers in developing countries. The RME model relies on a reward system based upon the intensity of contact between extensionists and farmers. Also the Netherlands Ministry of Foreign Affairs has a Special Programme on Biotechnology and Development Cooperation which has been exploring pilot projects on 'appropriate biotechnologies' which might better meet the needs of small-scale farmers in developing countries. A key feature of such approaches have been farmer participatory needs assessments to determine research priorities prior to initiation of research and development. However, most plant biotechnology research is conducted far 'upstream' of such 'downstream' structural changes in the agricultural extension and marketing sectors.

There has been a lack of biotechnology research which would enable key agricultural 'processes' at the on farm level which could improve or 'empower' farmers livelihoods. Yet in theory at least plant biotechnology research could be applied towards such goals, especially if there were better linkages between farmer participatory researchers/extensionists and plant biotechnologists. The CGIAR's Systemwide Programme on Participatory Research and Gender Analysis (SWP-PRGA) is currently exploring whether some biotechnologies might have utility in 'empowerment'-oriented farmer participatory plant breeding. The Centre for the Application of Molecular Biology to International Agriculture (CAMBIA) in Australia has for some years been trying to develop plant biotechnology tools which could empower low technology approaches to crop improvement in developing countries.

Some developing countries have had considerable success in establishing significant capacity in biotechnologies such as plant tissue culture and micropropagation and disease diagnostics, and in meeting farmers needs with such technologies. However, the strengthening of capacity in the plant molecular biotechnology research has proved more difficult to achieve in the short term, especially in a manner which is targeted to meeting country specific needs. In most instances, the existence of a conventional plant breeding programme which is operational is a necessary prerequisite for any rational application of advanced plant molecular biotechnology techniques such as marker assisted selection or transgenesis.

Despite some successes, there is a growing consensus that many international project based initiatives to transfer biotechnology capacity to developing countries have not been as successful as originally envisaged. The majority of developing counties have limited practical access to the tools and germplasm necessary to apply high technology biotechnology research to their national needs. The barriers to such access are many and mainly include lack of financial, scientific and infrastructural resources.

There are major differences between countries in their agricultural biotechnology capacity which would preclude any generalizations regarding the appropriateness of some biotechnologies for developing countries as a generic group. For instance, a number of developing countries in Asia and Latin America such as Brazil, China, India, Malaysia, Thailand, Philippines, and Indonesia have a relatively high level of plant biotechnology capacity, especially in early generation biotechnology areas such as plant micropropagation, transgenics, and marker assisted breeding.

In Africa, a significant number of countries have some limited biotechnology capacity in the areas of plant tissues culture and micropropagation. In some African countries, basic infrastructure and facilities for even the simplest plant tissue culture or micropropagation are not available. However, other African countries such as Morocco, Tunisia, Nigeria, Kenya, and Zimbabwe have some additional but limited capacity in plant molecular biology. While countries such as South Africa, Nigeria and Egypt have the capacity to generate transgenic plants, mechanisms to ensure that the plants can reach the end user i.e. the farmer are in many instances lacking.

The OECD's Development Centre has published an excellent study on the incentives, constraints and country experiences for integrating biotechnology in agriculture in different developing country situations. ISNAR's Intermediary Biotechnology Service has also produced a series of research reports which provide useful frameworks for decision making regarding national biotechnology priorities, planning and policies, based on the experience selected developing countries have had to date with the integration of biotechnologies into their agricultural research systems. The OECD study concluded inter alia that biotechnology research has not been closely integrated with the problems and constraints confronting the agriculture sector, nor with the obstacles to widespread diffusion of useful new biotechnologies, particularly to low income farmers. A lack of clear priorities and focus was identified. The OECD study called for reflection on the part of developing countries, scientists, NGOs, donors and the CGIAR on the development of innovative public/private mechanisms for the transfer of "public good" biotechnologies in developing country agriculture.

A 1994 survey of 45 organizations involved in the transfer of agricultural biotechnology revealed that most initiatives concentrate on the few developing countries with relatively advanced scientific and technological capabilities, and that developing country scientists and administrators are not always directly involved in their planning and design. Also, a 'brain drain' exists for many developing countries and regions whereby many of their scientists have moved to work/train in advanced biotechnology laboratories in the USA, Europe, Australia and Japan. If such scientists return to situations where there is no/little conventional plant breeding activity or infrastructure, the comparative advantage that they have learnt in plant biotechnology cannot easily be applied to the improvement of agriculture in their own countries.

There are currently few financial or other incentives for such scientists to return to their countries of origin. In the absence of public sector funding for such scientists upon return, it is likely that many such scientists will become technology adapters, and/or marketing agents for imported proprietary products/germplasm developed by non-domestic companies. In all countries, there is also a need to more actively involve end-users such as farmers and producers organizations in priority setting regarding the objectives of publicly funded agricultural biotechnology. At the international level, the International Federation of Agricultural Producers (IFAP) - an organization which represents a large proportion of the world's farmers - recently made a significant policy statement on 'Farmers and Biotechnology' at the 1998 World Farmers Congress. The IFAP policy statement raised issues regarding:

(a) the potential benefits of GMOs to different stakeholders,

(b) concerns of different stakeholders regarding GMOs;

(c) promoting freedom of farmers to operate

(d) promoting safety and accurate information

(e) increasing public sector research investment

(f) intellectual property rights

(g) addressing the needs of developing countries and

(h) maintaining biodiversity.

Scientific Developments in Agricultural Biotechnology

Agricultural biotechnology research is now generating a wealth of information of utility to crop improvement strategies. Plant biotechnology research is most powerful when it is integrated with conventional breeding or crop improvement approaches. In many countries, plant biotechnology research is being integrated with conventional plant breeding approaches wherever the technical and financial resources are available for such integration. Among the scientific community it is often highlighted that plant biotechnologies could also generate environmental benefits, especially where renewable genetic inputs can be effectively used to substitute for agro-chemical inputs. A vast range of approaches for the improvement of agronomic traits are either under study or in early development phases. These include the improvement of important agronomic traits such as; yield; pest and pathogen resistance; tolerance to abiotic stresses such as acid soils, drought, salinity and cold; herbicide tolerance; shade and high density planting tolerance; water and nutrient use efficiency; reduced mycotoxin contamination; crop reproductive biology; and enhanced nutritional and product quality through influencing quantity and quality of oil, protein, carbohydrates, nutrients, and novel substances.

Agricultural Genomics

Genomics is a new term that is used to describe the development and application of large scale, high throughput and parallel processing approaches to the functional analysis of entire genomes. Genomics technologies can allow the identity and function of tens of thousands of different genes to be analyzed simultaneously or in parallel. The science of genomics has arisen because the speed and scale with which the genomes of economically important organisms can be sequenced and functionally analyzed is increasing at a rapid pace. It is claimed that with current state of the art technologies a consortium of the leading sequencing institutions could now feasibly sequence the human genome from start to finish within three years. Technology spillovers from such human genome sequencing programmes will result in high throughput DNA sequencing being increasingly applied to the genomes of commercially important organisms such as human pathogens or crops. Agricultural genomics research is currently underway for a range of important agricultural species. Large scale DNA sequencing of some crop and plant genomes is now at an advanced stage, especially for Arabidopsis and rice.

Expressed sequence tags (ESTs) are small fragments of genes that have been sequenced and which can act as unique 'bar-codes' for each particular gene in an organism. Large scale EST projects can rapidly generate unique ESTs for 90%+ of all the genes in an organism in a very rapid and cost efficient manner. For instance, ESTs which each uniquely identify more than 95% of the estimated 120,000 human genes are available on public databases on the Internet. Publically funded research has now identified ESTs representing the majority of genes in the Arabidopsis (International Arabidopsis Genome Initiative) and rice genomes. The majority of such ESTs have been placed in the public domain via their publication in publically accessible EST databases on the Internet. It is also known that there are several large-scale commercially funded EST projects for crops such as maize, and soybean. However, to date most of this EST data is not publically available. Agricultural genomics initiatives are both generating genetic markers and also identifying genes that can be used either for marker assisted breeding or the development of transgenics with improved agronomic properties.

The application of molecular markers to genetic linkage maps of a wide range of crops is allowing the identification of the chromosomal (physical) locations of genes for improving yield and other complex traits important to agriculture. The underlying genetics of a wide range of quantitative agronomic traits are being unravelled through the identification of quantitative trait loci (QTLs) using a combination of molecular markers and advanced statistical breeding procedures. An idea of the wide range of agronomic traits and corresponding QTLs currently under investigation or improvement can be obtained from browsing through the abstracts of the annual Plant and Animal Genome Conferences. Powerful 'functional genomics' systems to explore the function of genes through 'knockout' and 'monitoring' strategies have also been developed using existing transposon mutagenesis systems, such as those which have traditionally been widely used for the genetic analysis of maize.

The development of DNA 'chip' technology is also set to revolutionize the scale at which genetic experiments can be done and can potentially allow simultaneous and rapid analysis of all of the genes in any organism at any particular point in time or environment. If this technology becomes cost effective in the same manner which has led semiconductor technology to become widespread it will have a very significant impact on how genetic improvement of crops is conducted. There are some large scale publically funded efforts to develop 'genomics' technologies and tools which would be made available to the academic plant biology community with "no strings attached" regarding intellectual property. In the US the Arabidopsis Functional Genetics consortium is adopting this approach.

The potential value of agricultural genomics has been recognized by the private sector judging by its current level of investment. High levels of public sector funding has also been approved for agricultural genomics initiatives in a few countries. For instance, the US government has approved $40 million to fund genomics initiatives for crops of national importance, and the Japanese government has earmarked substantial financial support for a national Rice Genome Project. Some analysts consider that the USA's research capacity in agricultural genomics now outstrips that of all other countries or regions and that the technology gap is widening in this area. It is thought that the promotion of the French GenoPlante agricultural genomics consortium is one national initiative which is trying to bridge such gaps between regions.

The Indian government is currently assessing proposals to strengthen India's capacity to conduct agricultural genomics research. Scientists participating in some of the plant genomics initiatives have stated that publicly available research tools and data are a priority for any such initiatives if broader scientific, social and economic spillovers are to be promoted. It is felt by many in the scientific community that international cooperation is critical to such large scale genome projects, both because genome projects are typically too large for any one country and the information forthcoming should be of benefit to the world and not just the countries that do the work. International cooperation is hence seen to be essential to ensure that overly protectionist proprietary positions are not taken which would inhibit the availability of potentially useful public domain information.

Comparative Genomics

A major discovery which is accelerating genetic mapping research has been the discovery that the genomes of distantly related crop species may in fact be quite similar to each other in terms of gene order and structure. For instance, a gene on the chromosome of one grass species can be anticipated to be present in a predicted location on a specific chromosome of other grass family species. For this reason, many cereal geneticists now view the grass family as a single genetic system whereby genomic information from one cereal species can be used to understand much of the genetics of other cereal species.

Because it is cheaper and easier to determine the DNA sequence of species with smaller genomes a number of species are emerging as the 'anchor genomes' which are expected to act as the 'Rosetta Stones' for understanding the larger genomes of other related species. Hence, a number of major multi-partner efforts are underway to determine the sequence and structure of major cereal crop genomes which will provide valuable information for understanding the genomes of other cereals. Examples include the International Grass Genome Initiative, International Triticeae Mapping Initiative, the International Rice Genome Initiative, the Rice Genome Project (Japan) and the National Corn Genome Initiative (USA).

The entire Arabidopsis genome will be sequenced and publically available by the year 2004 and is already providing valuable information regarding the genetics of economically important crops both within the Brassicaceae and within many other tribes. The crop wild relatives stored in the world's genebanks are valued as a unique source of genetic variation, but they have rarely been used for the genetic improvement of quantitative traits. Indeed it is widely acknowledged that exotic germplasm such as crop wild relatives is infrequently used by breeders.

A development of major significance has been the recent development of a powerful molecular marker based methodology whereby quantitative trait loci (QTLs) conferring complex traits such as yield and organ size can now be effectively transferred from wild relatives into crop varieties. This powerful technique has now been demonstrated for rice and tomato and is being rapidly applied to other crop species for which suitable molecular maps are available. This discovery suggests that the innovative use of molecular maps and markers will increase the accessibility and realisable value of wild and exotic germplasm.

Plant Micropropagation Technology

The rapid propagation of many desirable plant varietal genotypes using plant micropropagation technology is a relatively low technology 'appropriate' biotechnology which is now delivering tangible benefits to many farmers in both developed and developing countries. In addition to its rapid propagation advantages, such tissue culture can also be used to generate disease free planting materials. There are numerous examples of micropropagation initiatives which are delivering disease free planting materials to poorer farmers. These include local level micropropagation work on taro in Samoa, Musa spp and multipurpose trees in Kenya, potato in Vietnam and cassava in Colombia. In some instances there have been efforts to involve farmers organizations in the design and running of such micropropagation. This is recently the case for CIAT's proposed small scale cassava micropropagation work with the Colombian farmers' organization FIDAR and for ongoing potato micropropagation in Dalat province in Vietnam.

Micropropagation techniques have now been developed for a wide range of crops. for instance, China now has developed micropropagation technology for more than 100 crop species. In the Guangdong province, 3-4 million micropropagated banana plantlets are produced annually and 1 million exported. Micropropagated bananas are reported to have been successfully been adopted by poorer farmers in Guangxi and Guangdong in China. In 1994 it was estimated that the farmers in Guangxi received an extra income of about $723,000 as a result of adoption of approx. 600,000 disease free plantlets. Similarly, ten percent of the area of China planted to potatoes was derived from micropropagated virus free material in the early 1990s and yields are reported to have increased by up to 100-200%. As part of the FAO FARM project for farmers in rain-fed areas, China has developed micropropagation systems for multiplication of disease free bananas, potatoes, citrus and apple varieties, which are shown to be benefiting the target farmers.

Micropropagation of both food and export crops is also now routine in many Latin American and Caribbean countries. Large scale micropropagation is conducted for crops such as coffee, banana, plantain, taro, cocoa, cocoyam, sweet potato, apple, blueberry, raspberry, pineapple, citrus, grapes, papaya, mango, guava, potato, kiwi, cherry, pear, ornamentals, yams. The existing and potential benefits of adoption of such micropropagated plants by farmers have been substantial as testified by the following selected examples:

- In Mexico about 2.6 million people are dependant on coffee cultivation and production, However coffee rust affects 90% of coffee plantations. The Mexican government is facilitating the distribution of disease free plantlets both to protect labour in this area and export earnings.
- Costa Rica is the world's second largest banana producer. The national banana corporation, Corbana, a semi-public company is undertaking large scale micropropagation of nematode-free plantlets. The combination of a fallow period with nematode-free plantlets has eliminated the need for the use of nematicides in Corbana's plantations since 1987.
- At the Federal University of Rio de Janeiro, Brazil meristem culture of sweet potato cultivars to eliminate

viruses and pathogens, followed by field trials around Rio de Janeiro resulted the raising of yields from 9 to 19 tonnes.

- In Argentina, a company called Tecnoplant SA provides a service whereby it will 'clean' varieties delivered by customers from viruses and other pathogens, through tissue culture methods. It does this under contract for large-scale producers and planters. With appropriate subsidies such initiatives could be extended to producing disease free planting material of the locally adapted varieties of low income farmers.
- At the Biotechnology Institute, Santa Clara, Cuba clonal propagation of banana following in vitro micropropagation produces about 5 million plantlets annually. This is done using relatively low tech facilities at 25-30ºC with the result that yield from tissue culture derived plantlets was 30% higher than from conventional planting material.

Micropropagation capacity is less well developed in most African countries, yet it represents a technology which if better integrated with ongoing efforts in seed/planting material production and supply, could yield significant agronomic benefits to farmers. In particular, there are few links between genetic resources conservation initiatives and micropropagation initiatives for the more rapid supply of a wide range of healthy planting materials to farmers. There is no doubt that a strengthening of plant tissue culture and micropropagation capacity if effectively coupled to local seed/planting material delivery channels could generate massive benefits for many resource poor farmers. The application of such biotechnologies to local varieties and landraces of root and tuber crops could generate disease free plantlets while helping to both boost yields and also limit 'genetic erosion'. Apomixis is a naturally occurring phenomenon whereby some plant species can produce seeds without fertilization. While apomixis has been described in over 400 different plant species it is only found in a few crop species. The harnessing of apomixis genetics for heterosis breeding and general crop improvement could have significant implications for agricultural research.

One potential benefit is that it may be possible to develop true breeding hybrids which retain their yield advantages over generations. Indeed, the long lists of potential agronomic benefits that could be derived from apomictic systems which are easy to use in crop improvement suggests that apomixis may be one of the most important targets for concerted international research efforts. Such benefits include:

- use of hybrid vigour in almost every crop species, including the numerous crops that that never had hybrid technology available.
- survival, and immediate fixation of genetic combinations - including those made by widecrosses
- that are unfit under sexual reproduction, greatly expanding the diversity of utilized genetic resources.
- propagation through seed of crops that are currently vegetatively propagated such as cassava, potato, sweet potato and yams.
- plant breeding to become more simple - no need for inbred line production, nor use of male sterility, restorer lines, isolation plots - and therefore more responsive to environmental, economic and social changes.
- use of hybrid vigor technology at the level of the individual farmer.

Without changing his/her traditional practices, i.e. leaving farmers to select the best ears specifically adapted to their particular farming conditions to make the seed for the next cycle, large leaps in crop production might be achieved. In theory apomixis could allow both plant breeders to genetically adapt plants to specific micro-environments, rather than the current practice of adapting the overall cultivation environment to the crop plants requirements. Hence, through the use of apomixis plant breeding could become extremely rapid and responsive to specific micro-environments, cropping conditions and markets. This in turn could stimulate diverse strategies for more sustainable agro-ecosystem management and could have profound implications for biological resource management within agricultural systems.

The development of apomictic technology is one biotechnology that could provide major food and livelihood security benefits to farmers in developing countries. Considerable strides have been made in application of biotechnology for the development of apomictic crops and a number of promising research approaches are now underway. If any of the existing research approaches are successful, it is thought that apomictic crops may be developed within the next decade. However broad social benefits are only likely to occur if apomixis technology can be made accessible to developing countries and in particular resource poor scientists and farmers. Reflecting this concern, a consortium of the world's leading public sector laboratories which are attempting to develop apomictic technologies have expressed some concern that current IPR trends could hinder access to apomictic and other enabling biotechnologies for developing countries and poorer market sectors.

Integrated Pest Management (IPM)

Breeding for resistance to pests and diseases is a major ongoing activity for the vast majority of crops. There are a vast range of pests and pathogens for which conventional breeding has failed to produce crop varieties with durable resistance. There are also serious pests and pathogens which integrated pest management (IPM) approaches have not yet managed to address. A significant proportion of plant biotechnology is targeted at trying to develop options for control of pests and diseases for which there are currently few control options, although most such work is concentrated on pests and diseases of major commercial crops.

There have been many significant developments in elucidating the genetics of pest and disease resistance. The genetic basis of gene for gene interactions in plant pathogen interactions is being rapidly elucidated. Plant biotechnology research is identifying and isolating a vast range of genes and QTLs conditioning resistance against a wide range of pests and pathogens. Powerful techniques have been developed for 'scanning' crop genomes for related resistance genes and subsequent cloning of such genes. The effective transfer of resistance genes from one crop species to another has been demonstrated as a new option open to resistance breeding. Some transgenic approaches have demonstrating higher levels of broad spectrum resistance against pathogenic viruses, bacteria and fungi.

Plant biotechnology could potentially have significant substitution effects in the global insecticide market. The global insecticide market is estimated to be approximately US$ 8100 million per annum. The development of genes which can substitute for some of this need for application of insecticides could have a significant impact both on the environment and insecticide sales. The source of many such genes has been the bacterium Bacillus thuriengiensis (Bt) although many other organisms are now being screened for useful genes. Bt has been used for decades as an insecticide spray but has had limited use outside of 'organic agriculture' and Canadian forestry, in total accounting for less than 1% of the insecticide market. It is now estimated that US$ 2700 million of chemical insecticide applications could be replaced with Bt biotechnology applications, either as improved sprays or through expression in transgenic crops.

Bt transgenic crops are expected to be planted on an estimated 20 millions acres in the USA In 1997. Interestingly, it has required the advent of transgenic single gene resistance approaches to warrant legal requirements on farmers to engage in large scale resistance gene 'deployment' or 'recycling'. The planting of large contiguous crop areas to varieties which are relying on monogenes to condition resistance to important pathogens has long been recognized to be unwise because some monogenes can exert a selection pressure for resistance breaking strains of the pathogen to evolve. However there have been few public policy instruments developed to encourage that low selection pressures on pathogen populations are maintained in agricultural production systems.

The first generation of transgenic plants expressing the insecticidal Bacillus thuriengiensis (Bt) protein now require a resistance gene 'deployment' strategy to limit selection for an insect population that is insensitive to the particular Bt protein. Because Bt transgenic seed is proprietary it can be sold with a legal requirement that a certain % of the cropping area (i.e. refugia or mixed plantings) be planted to varieties which lack Bt transgenes to limit selection for Bt insensitive pests. Also Bt gene 'recycling' strategies are being established whereby a range of transgenic varieties each containing a different type of BT protein are released over time or geography to combat the potential ability of the insect pest to overcome any particular Bt resistance gene, and hence minimize the conventional 'boom-bust' cycles in variety-pathogen co-evolution.

Unless consideration is given to how monogenic resistance genes (whether transgenic or not) are spatially and temporally deployed against important plant pests and pathogens (especially airborne fungi) it is likely that the familiar 'boom-bust' cycle will be perpetuated and valuable genetic resources will have been wasted. It is suggested that multiline and mixture strategies for resistance (trans)gene deployment nay become more feasible because of the increasing number of resistance genes being isolated and the ease with which transgenic crops can be established.

Past failures with multiline and mixture strategies may have been partly due to a lack of control by breeders over the accurate deployment of resistance genes by farmers and that stronger property rights (IPRs or user contracts) over varieties or resistance genes would allow a requirement to be placed on the variety user to ensure that optimal gene deployment or recycling practices are followed. Less coercive resistance gene deployment strategies are reportedly under development or in use in some IPM programmes, including the FAO's Farmer Field School Approach. However, the lack of fruitful interaction between farmer participatory IPM approaches and transgenic or biotechnological approaches to pest/disease management has been noted. It is suggested that there is a need for more collaborative interaction between experimenting farmers and scientists, especially regarding pest and disease problems that cannot be solved by local level research alone.

While dangerous pesticides have been over-used in many developing countries, herbicide use has been very low even thought most herbicides are far less toxic than pesticides. Yet weed control is a major time, labour and resource consuming task for most farmers, and in particular for resource poor farmers with limited access to inputs such as affordable herbicides etc. It is estimated that more than 60% of developing country farmers time is spent in weeding. Much of this weeding is often done by women and children and is often unpaid work. There is a very convincing case to be made that herbicide resistant crops could in the near term offer significant, economically accessible, advantages to many farmers in developing countries, in particular poorer farmers with limited labour availability. In particular, there are many weed problems faced by farmers in developing countries for which no effective control measures have been developed, with or without herbicides. These include the parasitic broomrapes and witchweeds.

The areas infested with such weeds are vast and expanding. For example, a survey of 180,00 square kilometres in Nigeria found that 70% of fields were infested with witchweed seeds. In the seven agro-ecological zones of sub-Saharan Africa witchweeds are generally listed as the worst pests affecting agriculture. Witchweeds infest the grain crops of more than 100 million people in sub-Saharan Africa and Asia, reducing yields by 50%, and by more in drought years. Labour intensive weeding is largely ineffective against such weeds. Crop yields could potentially be doubled if such weeds could be controlled. In addition labour spent in weeding could be released for other more productive activities, such as increasing literacy and schooling for children. It was recently found that it is possible to control Striga spp using imadizoline resistant maize. One such strategy of potential utility to developing country farmers is being developed whereby only the transgenic seed is treated with high levels of a systemic imadizoline with a resultant excellent control of Striga. Using $5 of herbicide gave $100 of increased maize yield per hectare in Striga - infested areas in Kenya. Such strategies would require resistance management measures to ensure that Striga resistance to the herbicide would not evolve.

Herbicide resistant crops could form part of integrated weed management systems, where resistance management strategies are used to ensure that herbicide tolerance does not develop in the weed flora. However it would seem that current biosafety regulations will limit African farmers access to herbicide tolerant crops in the near term, even for crops such as maize which have no weedy wild relatives in Africa. Hidden hunger, such as protein and micronutrient deficiencies, is a widespread and endemic problem for the worlds poorest people, especially women and children.

A range of transgenic approaches have now been developed to nutritionally improve the amino acid profile of crop protein, either by transferring genes encoding more nutritious proteins from other species, or by manipulation of crop biosynthetic pathways to increase the nutritional profile of endogenous proteins. Where transformation protocols have been developed, many important legumes such as peanut, beans, clover etc could feasibly be nutritionally improved through the transfer of methionine-rich protein genes from species such as sunflower. Sunflower was chosen as the gene source because, unlike Brazil nut, the seed is not known to cause any allergic reactions. Insufficient intake of dietary vitamin A is implicated in the death of approximately 1-2 million children annually. In South-East Asia, every year an estimated 5 million children develop the eye disease xerothalmia. Unfortunately, many staple crops such as rice are deficient in dietary vitamin A. In addition, the vitamin A containing tissues of rice are removed during milling of rice.

Genetic engineering have now developed milled rice which accumulates vitamin A to provide one means of facilitating increased dietary intake of vitamin A from staple foods. If technology transfer of such vitamin A rich crops can reach its intended clients, it is likely that the transgenes used to increase vitamin A production could be applied to other crop species or varieties in locations where vitamin A deficiency is a medical problem. Similar approaches to combating micronutrient deficiencies by increasing the both content and availability of iron in transgenic rice are showing much promise.

The biosynthetic pathways which produce commercially interesting compounds in plants and other species can now be manipulated by 'metabolic engineering' of the pathway so that higher levels of desirable compounds may be obtained. Seed oil content has been increased using this strategy. Such strategies are currently the focus of much commercial biotechnology interest. It is now becoming possible to tailor the specifications for the modification of vegetable oils in transgenic plants that more specifically meet end user needs. One example is the production of higher levels of laurate in rapeseed. Metabolic engineering approaches are also being explored for the transfer of the key biochemical components of C4 photosynthesis to those crops which rely on the less energy efficient C3 photosynthesis.

Similar antisense or gene-suppression technologies approaches can be used to 'switch off' or control the timing of the production of undesirable (or desirable) compounds. For instance, the Flavr Savr tomato developed by Calgene uses anti-sense technology to suppress one of the genes responsible for ripening, so that tomatoes remain on the vine longer and become sweeter without going soft. The introduction of genes which delay ripening or spoilage could help to reduce the post-harvest losses of some perishable vegetables and fruits, especially in situations where poor farm-to-market roads, inadequate transportation, and inadequate storage facilities exacerbate post-harvest losses. Similar types of transgenic approaches are being extended from tobacco to other crops such cassava in order to inhibit leaf senescence.

A number of transgenic methods have been developed for changing the molecular structure of plant structural or storage compounds so that the crops are more digestible for either humans or domestic animals. For instance, lignin is a plant compound that adversely affects pulp and paper production processes and which also lowers the nutritional value of animal feeds. Mutant lines of maize, sorghum and pearl millet have been described that have a reduced lignin content and improved digestibility.

Human disease is a major constraining factor to labour availability in many agricultural projects and to socio-economic development general in developing countries. Lack of effective cold storage facilities limits the efficacy of linear supply chains for many vaccines. Production of effective oral vaccines against major tropical diseases in transgenic plants may be an extremely appropriate and low technology means of decentralizing both vaccine production and distribution in developing countries. The potential feasibility of producing oral vaccines in transgenic plants has now been demonstrated for diseases such as cholera - toxin B1 and hepatitis B1. If they are made widely accessible, such transgenic plants may be of major utility to hospitals and medical centres in providing a reliable and cost effective supply of heat-stable vaccines and other protein based pharmaceuticals.

Risks and Benefits of Modern Biotechnology

In considering the potential risks and benefits of modern biotechnology, it is useful to distinguish technology-inherent and technology-transcending risks. Technology-inherent risks are those where the technology itself has potential risks to human health, ecology, and the environment. Technology-transcending risks include those that are not specific to the technology but where its use may have risks. For biotechnology these include the risk of increasing the poverty gap within and between societies, reducing biodiversity, and antitrust and international trade issues.

Health Risks

Potential health risks of genetically improved organisms relate to assessing and minimizing the risk of food allergens in genetically improved food. New biotechnology based methods allow the identification, characterization, and minimization of risks of food allergens. Genetically improved crops and food, and the risk of allergens associated with them, are now a concern throughout the world, especially in industrial countries. More than 90 percent of food allergens that occur in 2 percent of adults and 4-6 percent of children are associated with eight food groups. Allergenicity of genetically improved foods can be raised in crops and foods either by raising the level of endogenous allergen or by introducing a new allergen.

Assessment of the risk of allergens is a challenge. The International Life Sciences Institute (ILSI) has developed a decision tree that provides a framework for risk assessment. It uses the following criterion: that an introduced protein in a food is not a concern if there is:

(1) no history of common allergenicity,

(2) no similar amino acid sequence to known allergens,

(3) rapid digestion of the protein, and

(4) the protein is expressed at low levels.

Protocols enable assembly of the data to judge food against this criterion. It is also important to inform consumers of any potential risk. A key concern of consumers is being able to identify where allergens are found. Consumers want to know where the potential for food allergens exists. Any protein added to food should be assessed for potential allergenicity, whether it is added by genetic engineering or by manufacturing.

There are several related areas of concern with regard to potential human health risks of genetically improved foods: toxicity, carcinogenicity, food intolerances; the risk of the use of gene markers for antibiotic resistance; other macromolecules aside from protein that could be potential allergens; and nutritional value. Methods of testing and evaluating risks of toxicity and carcinogenicity are well established for food. The question remains as to whether developing countries can implement and use currently available technologies and protocols to assess food allergens and other health risks. The techniques are well established, and should be readily implementable by trained professionals. Although no clear cases of harmful effects on human health have been documented from new genetically improved food, that does not mean that risks do not exist and they should be assessed on a case by case basis.

Environmental Risks

The risks policymakers and regulators need to assess include the potential for spread of traits from genetically improved plants to the same or related species, plants, the buildup of resistance in insect populations, and the potential threat to biodiversity posed by widespread monoculture of genetically improved crops.

- A transparent, science-based framework is required, which assesses risks on a case by case basis and takes account of all stakeholder views.
- Environment-related issues to be considered in each case include the possibilities for gene transfer, weediness, specific trait effects, genetic and phenotypic variability, and expression of pathogenic genes.
- Risk management needs to consider the prospects for managing any specific risks identified with a proposed release
- Experience is accumulating in the management of the Bt genes in transgenic cotton varieties in several countries and this needs to be closely monitored.
- An agricultural sustainability protocol that balances risks and benefits may have value for the approval and use of new crop varieties.

The risks lie in the management of the cropping system involving soil, water and other inputs. There is a need for the establishment of baseline information in the environment where such introductions are to be made. There is very little known on this, although some understanding has been gained over recent years, and further R&D is required. The information derived from such an assessment needs to be handled through risk management associated with "plants as plants." Risk management involves the consideration of traditional cultural practices that have evolved over time, and new knowledge gained from research in agronomy, plant pathology, entomology, weed science, plant biology, soils, microbiology, and other disciplines.

Risks to Biodiversity

Risks to biodiversity and wildlife are important issues in particular environments. Careful assessment is necessary of the risks associated with the possible creation of new selection pressures coming from the introduction of genetically improved organisms into the environment. Of special concern is the potential impact on biodiversity of genetically improved organisms as the selection pressures wield influence in the species composition of the ecosystem. These concerns merit further study, especially on the behavior of genetically improved organisms in the open environment. The framework for strategic planning in the deployment of genetically improved organisms should be formulated with sustain-ability as the primary concern.Both food safety and biosafety regulations should reflect international agreements and best practice and a given society's acceptable risk levels, including the risks associated with not using biotechnology to achieve desired goals.

The principles and practices for assessing the risks on a case-by-case basis are well established in most Organization for Economic Cooperation and Development (OECD) countries and several emerging economies. These principles and practices have been summarized in a series of OECD reports published over the past decade or more. National, regional, and international guidelines for risk assessment and risk management provide a basis for national regulatory systems. Biosafcty guidelines are available from several international organizations including the OECD, United Nations Environment Program, United Nations Industrial Development Organization, and the World Bank.

Regulatory trends to govern the safe use of biotechnology to date, include undertaking scientifically based, case-by-case, hazard identi-fication and risk assessments; regulating the end product rather than the production process itself; developing a regulatory framework that builds on existing institutions rather than establishing new ones; and building in flexibility to reduce regulation of products after they have been demonstrated to be of low risk. Biotechnology is not inherently different to other technologies with respect to economic and social impacts, as long as it focuses on the problems that affect poor people. One important difference is that research on biotechnology has largely taken place in the private sector with proprietary technologies and an orientation to commercial agriculture. This implies the need for a strong role for the public sector, including increased resources, to address developing country priorities.

Socioeconomic Risks

There is a risk that modern science may bypass the needs of poor people. Biotechnology is only one tool in addressing the challenges of food security and poverty. There is a need for biotechnology to be integrated with appropriate policies and other conventional R&D programs. The positive and negative impacts of biotechnology should be monitored over time in terms of who and what are affected and how they are affected. Monitoring impact will provide guidance for public policymakers in the future.

Unless countries have policies in place to ensure that small farmers have access to delivery systems, extension services, productive resources, markets, and infrastructure, there is a risk that the introduction of agricultural biotechnology could lead to increased inequality of income and wealth. In such cases, larger farmers are likely to capture most of the benefits through early adoption of the technology, expanded production, and reduced unit costs.

Biotechnology has potential to reduce input use, reduce risk to biotic and abiotic stress, increase yields, and enhance quality-all traits which should enable the development of new crop varieties that are appropriate to poor producers and consumers. Modern biotechnology is not a silver bullet for achieving food security, but, used in conjunction with other agricultural research, it may be a powerful tool in the fight against poverty. It has the potential to help enhance agricultural productivity in developing countries in a way that further reduces poverty, improves food security and nutrition, and promotes sustainable use of natural resources. Solutions to the problems facing small farmers in developing countries could benefit both farmers and consumers.

The benefits and risks need to be assessed on a case by case basis, weighing the risks and benefits for each particular situation. The benefits of new genetically improved food to consumers are likely to vary according to how they earn their income and how much of their income they spend on food. Consumers outnumber farmers by a factor of more than 20 in the European Union, and Europeans spend only a tiny fraction of their incomes on food.

Similarly, in the United States, farms account for less than 2 percent of all households, and the average consumer spends less than 12 percent of income on food. In the industrial countries, consumers can afford to pay more for food, increase subsidies to agriculture, and give up opportunities for bettertasting and better-looking food. In developing countries, poor consumers depend heavily on agriculture for their livelihoods and spend the bulk of their income on food. Strong opposition to genetically improved foods in the European Union has resulted in restrictions on modern agricultural biotechnology in some countries. The opposition is driven in part by perceived lack of consumer benefits, uncertainty about possible negative health and environmental effects, widespread perception that a few large corporations will be the primary beneficiaries, and ethical concerns.

Future of Agriculture

Over the past decade there has been some advocacy of a need for a pro-poor bias in the development and dissemination of modern biotechnologies. However, whether any benefits of current plant biotechnology research will actually reach poorer farmers and consumers without major public sector intervention is an open question. There are many such poorer people. Over 1100 million farmers, in many different farming systems and environments, are economically active in agricultural production globally.

The vast majority of the world's farmers are known to have a limited level of access to external inputs or other productive resources. Resource poor farmers, by definition, are unlikely to have easy financial access to agricultural inputs such as pesticides, fertilizers or irrigation. Moreover, it is now thought that an increasing majority of the worlds resource poor farmers are women. For instance, over 70% of the people in developing countries living below the poverty line are women, the majority of whom live in rural areas. While such resource-poor farmers practice approximately 60% of global agriculture, they produce 15-20% of the world's food. However, when looked at another way the small-scale resource poor farming sector is responsible for 80% of agricultural production in developing countries and is key to future food security.

The low productivity of resource poor farmers tends to perpetuate rural poverty to the extent that of the more that 2,500 billion people in developing countries who live in rural areas, approximately 1000 million live below the poverty line: 633 million in Asia; 204 million in Africa; 27 million in the Near East and North Africa; and 76 million in Latin America. Science alone is unlikely to provide a 'technical fix' for alleviating such poverty. There are many processes, factors and socio-economic structures underlying rural peoples poverty such as; lack of access to land and other productive resources, low purchasing power, political powerlessness, fragile environments, peripherality from markets, etc. In this milieu, agricultural research is but one factor which could have differential impacts on rural poverty.

Indeed, the potential contribution of biotechnology to developing country agriculture or to poverty alleviation is considered to have been overstated, in the short term at least. Yet over the longer term there is little doubt that some biotechnological approaches to crop improvement could generate social, economic and environmental benefits if specifically targeted at specific needs, especially those of poorer groups. Such needs might for instance include the reduction in pesticide use via insect/disease resistant crops, improved nutritional composition of crops, elimination of toxic substances or allergens, developing early maturing varieties, reducing post-harvest storage losses, abiotic stress tolerant crops or reducing labour demands at appropriate times during the cropping cycle. For commercial reasons, richer farmers are likely to be the main target market for most privately funded plant biotechnology research.

The many resource poor farmers in developing countries who depend on an income of less than a $1 per day are not likely to be a near term target market for most of the agricultural biotechnology companies. If plant biotechnology research is to be better targeted to addressing the needs of poorer farmers, it will be necessary for relevant public sector institutions to more transparently identify which farmers needs are of concern to their research or funding agenda. Yet, even in the face of increasing population pressure and such large numbers of resource poor farmers in developing countries, internationally there are only a handful of underfunded plant biotechnology initiatives with an explicit focus on poorer farmers as their primary clients/markets. This may reflect to overall current bias in agricultural biotechnology research to commercial rather than social markets. The major agricultural biotechnology companies are concentrating on the development of two broads types of proprietary traits;

(i) input traits such as herbicide tolerance, insect or disease resistance etc and

(ii) output traits which improve the nutritional contents of foods or exhibit unique properties for very specific end uses or markets.

Such input and output traits will be incorporated into existing elite varieties to provide higher value seed with further added value, which may offer to the farmer lower costs or higher yield, and increased value of the end product. Initially, it is likely that biotechnology generated varieties brought to the market will focus on input traits. However, the long-term commercial potential of plant biotechnology is considered to be in the development of value-added output traits that will address a wide range of specific needs or market niches.

To assess the level of demand for particular types traits or products, the marketing departments of most agricultural biotechnology companies typically conduct surveys of farmers with different income levels to determine what commercial products might be developed by their technologists to meet the needs of customers who can express their demand in financial terms. Because demand driven companies are likely to make more money, most marketing departments in companies are responsive to customer concerns and demands. However, within many public sector research organizations, there is often an absence of demand driven biotechnology research agendas, especially in relation to the agronomic or socioeconomic needs of poorer farmers or developing countries. Some publicly funded plant biotechnology research could in effect be competing with the private sector for the same customers or clients. For the private sector, poorer farmers and consumers are by definition not a lucrative market and are unlikely to exert any effective 'demand pull' on the private sector research agenda.

In theory, the onus would therefore fall on the public sector to fund and perform any research required to meet the differential needs of poorer farmers or consumers. Yet, in most public sector institutions or funding bodies there are currently few priority setting or needs assessment mechanisms in place to help guide the direction of publicly funded plant biotechnology or crop improvement research towards meeting the immediate needs of poorer farmers.

In some instances plant biotechnology may represent one of a number of competing technological approaches to addressing a particular agronomic problem. For instance, a particular pest problem might equally be addressed through conventional plant breeding, through a transgenic approach, or through an integrated pest management (IPM) approach or any combination of these. Within the broader agricultural research community there is often a lack of priority setting mechanisms and relative cost-benefit analyses to determine which available technological approaches may be the most suitable within particular timeframes for addressing prioritized agronomic needs or the needs of particular groups of farmers.

2

Crop Production

Biotechnology is a word used to describe the process of using living organisms or any part of these organisms to create new or improved products. The term is most often associated with methods and techniques developed during the past 20 years that make use of the cells and biological molecules of living organisms. But biotechnology, in some ways, is not very new at all. For six thousand years, we have been using microbes to make useful products such as bread, yogurt, and cheese. Also, for thousands of years, people have altered the genetic makeup of plants and animals to make them more useful.

Growers have improved the characteristics of crops by planting seeds selected from the biggest and best individual plants. Through this selection process, desirable characteristics have been promoted. Plants have been cultivated and animals have been domesticated, changing them from their wild beginnings to forms more useful to humans. While this process is slow, it has been quite successful. Because people involved in agriculture could clearly see the benefits of improving plants and animals, scientists developed a keen interest in understanding how organisms inherit certain characteristics. Plant breeding as a science began in the 19th century with the discovery of how plant traits are inherited.

Gregor Mendel discovered that the characteristics of plants and animals are determined by genes, which are a fundamental part of the nucleus of each cell in an organism, and that the genes carry certain traits from each parent to the offspring when organisms reproduce. Of course, not all plants and animals of a given species are identical; some have different traits that make them more desirable for agricultural production—for example, higher yield or greater resistance to drought. Plant breeders select plants with desirable traits after the exchange of genes by cross-fertilisation between two parents. They then ensure that these traits will be passed on to future generations, thus creating a new variety of the plant.

Most of the early breeding programs were based on the collection of different specimens of the same species of plant from throughout the world. This approach allowed breeders to obtain as much of the genetic variation as was available within the species worldwide. Then through conventional breeding techniques they tried to utilise desirable characteristics to improve the domesticated varieties of those species.

This process is the basis for the development of essentially all varieties of plants and animals used in agriculture today. However, it is slow, commonly requiring 10 years before a new variety is ready to be released. But the biggest problem is that a desirable characteristic being sought to improve a given species may not be found among any of the plants or animals of that species in the world. Further improvement by conventional breeding can thus reach a dead end for that desired trait. For example, if resistance to a particular insect is needed in a given crop species and no such resistance is found in any plant of that species in the world, then protection of that crop may be dependent upon insecticides. With new techniques, however, the path to genetic improvement need not end that soon. If resistance to the insect is found within a different species, modern methods make it possible to transfer that resistance to the plant species of interest. The same methods could be used to increase yield, improve cold hardiness, or add any desirable characteristic.

The source for genetic resources is expanded to include all living species, making the possibility of improvements virtually unlimited. Modern techniques fall into three categories: cross-breeding between species, cell fusion, and genetic engineering. The reproductive process through which genetic material is transferred between individuals of a species does not normally work between species. The desire to improve certain crops, however, has led plant breeders to develop processes for interspecific hybridisation to transfer genes from one plant to another plant of a related species. This transfer has been accomplished during the last 80 years only through the discovery of efficient ways to circumvent the natural barriers to the exchange of genetic information between species.

The processes for interspecific gene transfers are usually laborious, time consuming, and quite often unsuccessful. There are many steps in the process of developing a hybrid plant, and each step is a possible point for failure, either through death or sterility of the plant. Even when a hybrid plant is created, there is a possibility that undesirable traits that affect crop quality, yield, or adaptation to stress may be transferred along with the desirable trait. Even with all of these problems, interspecific hybridisation has been responsible for many improvements in the tolerance of crops to physical stresses, resistance to diseases and insects, and increased yield potential for certain crops. The process has even made possible the development of a new grain crop, triticale, from the hybridisation of wheat and rye.

Microbiologists and breeders have been looking for alternate means to transfer genes. Such exchange of genes depends on the scientist's ability to regenerate a plant from tissues or organs. Much effort has been devoted to cell fusion, a process for combining two cells of the same or different species in the presence of certain chemicals or electrical current. In addition to the problem of regenerating a complete plant from the resulting cell, it is often difficult to achieve a stable hybrid because of incompatibility between the parent species. As with interspecific hybridisation, many genes are transferred during the cell fusion process, and thus unwanted traits may be transferred with the desirable ones. Further crosses may be necessary to obtain plants with the desired combination of parental traits.

Genetic Engineering of Plants

Genetic engineering is the method of transferring genes is to move the DNA directly. Direct combining of DNA is called genetic engineering. New techniques make it possible to use genes from totally unrelated organisms. First, scientists identify genes that control specific functions or characteristics—for example, improved insect resistance or production of enzymes for the degradation of a certain herbicide. Then they move these genes into the host plant. With genetic engineering, the time from discovery to incorporation of a desirable trait into a host organism will usually be shorter than with the use of classical breeding techniques to produce new varieties.

Genetic engineering may help crop producers reduce dependence on pesticides by affecting the plant's resistance or tolerance to pests in the field. It can also affect production efficiency and profitability by improving the climatic adaptation of the plant and by enhancing the postharvest storability of the product. Crop breeders have traditionally considered the development of pest resistance in crops and animals to be of primary importance. Genetic engineering can speed up the process of developing resistant varieties and potentially can introduce genes that will provide new ways for plants to withstand pests.

There has been much interest in using genetic engineering for insect control. The interest in developing insect-resistant varieties results from the desire to reduce the extent of pesticide use and to avoid the development of pesticide resistance in some of the important insect pests. The development of crops with insect resistance is not a new idea. Classical breeding programs have been developing varieties with insect and disease resistance or tolerance for many years. The problem has been that insects are generally able to overcome the resistance in 2 to 10 years. Another aim of biotechnology in insect control is to provide materials that are selective, will not affect nontarget species, and to which insects will not easily develop resistance.

Much of the work to date has focused on applications of *Bacillus thuringiensis* (Bt), an insect control bacterium, but other approaches are being explored as well. Bt has been used as a biocontrol agent for 20 years. It is a naturally occurring bacterium that produces protein crystals that are toxic to insects. These proteins are very specific insecticides with activities against some species of caterpillars, mosquitoes, flies, and beetles. Recent research on Bt has followed two approaches. The first is the selection and development of Bt strains that are specific for other pests or that produce higher concentrations of the compounds. Different strains of Bt to control plant- and animal parasitic nematodes, animal-parasitic liver flukes, protozoan pathogens, and mites have been identified.The second approach is to move the gene that controls the production of the Bt proteins into crop plants.

Bt genes for control of an insect species have been incorporated into tomatoes, tobacco, cotton, and corn with good results. Bt proteins are highly specific to particular insect species. Therefore, to achieve control of several different insect species at one time, many different gene codes would have to be incorporated into the plant's genetic makeup. The incorporation of Bt genes into plants causes concern that insects will likely develop resistance to the Bt strain. The probability is especially high in current cases where the gene is expressed throughout the plant.

If three to four generations of an insect feed on the plant leaves or fruits throughout the growing season, the insect would be constantly exposed to the chemical. It is likely that the insect will develop resistance to the chemical. It may become possible to direct the expression of the Bt products to only the fruit tissues, meaning that only one generation of insects would be exposed. It may also become possible to incorporate the genes in such a way that they are expressed only after an insect begins to feed on the plant. Finally, several different Bt genes may be introduced at one time, making the development of resistance more difficult. Research has been conducted on developing insect resistance mechanisms that can be used in different plant species. This work has centered on a chemical complex known as the cowpea trypsin inhibitor (CpTI). Results obtained so far suggest that CpTI will control a wider range of insects than the specific Bt products. In experiments on tobacco, good control of foliage feeding caterpillars was achieved. There appears to be no adverse affect on humans because CpTI comes from the cowpea and has not appeared to cause health problems when cowpeas are eaten raw or cooked.

Another area of interest is the incorporation of insecticide resistance into the natural predators or parasites of major and secondary insect pests. Secondary insect pests are those that are controlled by natural predators until insecticides are applied to control a major insect pest. The insecticides also kill the natural predators, allowing the secondary insect pests to flourish. Resistant types of the predators and parasites are being sought so they can be used for control. Modern genetic techniques will help in understanding the mechanisms of insect resistance and in the development of greater resistance in the future.

Genetic Improvement and Herbicide Resistance

Herbicide tolerance is a plant's ability to endure the effects of a herbicide at the rate normally used in agricultural production. *Herbicide resistance* is the ability of a plant to be unaffected at any feasible rate of herbicide application. Most crops are resistant to one or more herbicides. For example, corn is naturally resistant to atrazine, corn and soybeans are tolerant to alachlor (Lasso) and metachlor (Dual), but soybeans are not tolerant to atrazine. Biotechnology has provided plant scientists with additional tools to determine the chemical and genetic modes of action of many of these herbicides and also the mechanisms that account for a plant's natural tolerance or resistance to herbicides. As a result, scientists will use this knowledge to incorporate herbicide tolerance into crop plant species.

Several different methods have been used successfully to develop herbicide-resistant crop varieties. One method is to find a closely related species that has herbicide tolerance or resistance and then, through classical breeding techniques, to incorporate that tolerance into the desired plant. This process has been successfully applied to canola using a related species that was found to be resistant to atrazine. This new atrazine-resistant variety of canola is currently being cultivated in Canada. A second method has been the use of cell or tissue culture to test many different lines of plants for tolerance to a specific herbicide. As a result, Pioneer Hi-Bred has been able to develop three corn hybrids for use with Pursuit, a herbicide that typically kills conventional corn hybrids. A third method has been to determine the specific gene or genes within a plant or microbe that allow tolerance or resistance to a specific herbicide. This gene is inserted into the plant of interest, which is then tested for tolerance to the herbicide. This method has made it possible to develop cotton tolerant to the herbicide bromoxynil and to soybeans tolerant of glyphosate.

Bioherbicides for Weed Control

Bioherbicides are fungal or bacterial products selected for their ability to cause disease in specific plants, such as weeds, without harming desirable plants. They may be applied in the same manner as conventional herbicides. At present, two bioherbicides are being marketed for the control of specific weeds that are normally hard to control: DeVine for control of strangler vine in Florida citrus and Collego for control of northern jointvetch in rice and soybeans in Arkansas, Louisiana, and Mississippi. DeVine has been so successful in destroying strangler vine that the market for the product has almost been lost. The reason for its great effectiveness is that the product remains in the soil and gives 95 to 100 percent control for 6 to 10 years after a single application.

There are other bioherbicides in various stages of research and development for such things as control of prickly sida in cotton and soybeans, control of sicklepod in cotton and soybeans, control of spurred anode in cotton, control of velvetleaf, and growth suppression of water hyacinth. Biotechnology will play a major role in helping overcome problems in manufacturing these bioherbicides by the development of better fermentation processes, as well as assisting in the isolation of the genetic determinants of virulence, specificity, sporulation capacity, toxin production, and tolerance to climatic stresses. A second application of biotechnology that shows promise is the development of microbial and secondary plant products for use as herbicides. Much effort has been devoted to determining the actual compounds associated with allelopathy. Many of these compounds have limited selectivity and a lack of stability. However, one herbicide derived from chemicals found in a naturally occurring microbe is being marketed in Japan. Herbiaceae exhibits strong herbicidal activity against a wide spectrum of grass and broadleaf weeds when it is applied to foliage.

A more important role for these compounds is to provide models for the development of new chemicals that could be produced as commercial herbicides. It may be possible to produce synthetic derivatives of these chemicals that are more stable under field conditions, have greater selectivity than the natural chemicals, or have other advantages over the original chemicals. Several companies have developed chemicals based on this natural herbicide chemistry. It is believed that the naturally occurring herbicides will be safer for the environment because many of them are degraded rapidly in the soil. Concerns have been raised about these developments. One is that development of herbicide-resistant or herbicide-tolerant crop varieties may lead to overdependence on herbicides. Many of the herbicides being used for the development of crop herbicide resistance are broad spectrum, low-mammalian-toxicity chemicals that are thought to be safer for the environment than conventional herbicides. However, attempts are also being made to develop varieties with resistance to certain other compounds, such as atrazine, that are more persistent and have been found in groundwater in certain areas of the United States.

Continued use of these older chemicals, ones that tend to persist in the soil and may move into groundwater, could cause environmental problems.

Associated with the exclusive use of one herbicide for weed control in each crop in a rotation is the possible development of herbicide-resistant weeds. There are already over 100 species of weed plants known to have developed resistance to one herbicide or another. Much of this resistance has resulted from the continual use of one type of herbicide increasing the probability of developing resistant plants within the weed population. The use of a comprehensive weed control program, including rotating chemicals based on their mode of action, should allow the successful use of the new, "safer" herbicides. Another concern is the possible movement of the "engineered" genes from the host plant to related weed species. This possibility has not received much attention in the United States because in most cases our weed species are not closely related to crop species. An exception is the vegetable industry in California because many of the vegetable crops have closely related weed relatives in the wild. On the other hand, the possibility of outcrossing may not be very likely, considering the lengths to which plant breeders had to go to accomplish interspecific hybridisation before genetic engineering was introduced.

Control of disease is a subject of great interest for biotechnologists. The majority of advances have been in control of viral diseases. Because most viruses are spread mechanically or through insect vectors, control efforts have traditionally revolved around control of the vectors and destruction of diseased plant material. Viruses are composed of two parts—the viral DNA and a coat protein that surrounds the viral DNA. Researchers have known about the phenomenon of cross-protection: that infection of a plant by a mild strain of virus can often protect the plant from a serious infection by a more virulent related strain.

The researchers have recently discovered that it is the presence of the coat protein that restricts the infection by the virus in cross-protection. By incorporating the genes for the coat protein into the plant it is possible to have the plant itself produce low levels of the coat protein. These low levels of the coat protein delay or restrict infection of the plant by the virus. An example is the incorporation of genes for the production of the coat protein of the tobacco mosaic virus (TMV) into tomato plants. Untreated tomato plants infested with TMV showed up to 60 percent loss in yield, whereas resistant plants showed no yield decrease after inoculation with the virus. The level of the viral coat protein in transgenic plants is lower than that found in plants infected with endemic strains of the virus.

Cross-protection via gene transfer offers a number of advantages. The first is that the protection is essentially permanent, similar to that afforded animals by vaccines. The need for the use of chemicals to control insect vectors is also reduced. Other possible aspects of disease control in which biotechnology may play a role include the use of plant disease biocontrol fungi and the development of fungal resistance in plants. Plant disease biocontrol fungi are naturally occurring organisms that are antagonists for certain soil-borne plant pathogens. Biotechnology will play a role in better understanding the mechanisms for host specificity and virulence of these fungi and may lead to the production and formulation of control products.

Improving Product Value

Traditional breeding has developed cultivars with high concentrations of lysine, an essential amino acid. The increase in lysine content, however, was accompanied by a 10 percent decrease in yield. Scientists are working to incorporate a gene in corn that will increase the amount of lysine without the associated yield decrease. They have also developed corn with a higher-than-usual oil content for use as a source of vegetable oil and feed. Oilseeds are another crop of interest. Researchers are seeking to improve the nutritional qualities of vegetable oils along with the type and concentration of these oils within the plant seeds.

Work is under way to develop varieties of sunflowers and canola with high oleic acid content. Researchers have also produced a low-palmitic-acid soybean, making the oil lower in saturated fats and therefore more comparable nutritionally to canola oil. Bioengineers have recently been able to alter canola so that it can also produce lauric acid, a key raw material for the soap, detergent, oleochemical, personal care, and food industries. Currently, the only commercial sources of lauric acid are coconut and palm kernel oils.

Food producers and processors are interested in the ripening of fruit and in the processing quality of fruits and vegetables. An example is improving the processing quality of tomatoes, a major food crop and a significant source of vitamin C. The characteristics of most interest to processors are percentage of solids and consistency. It has been estimated that an increase of 1 percent in tomato solids would save the processing industry between $70 and $100 million annually. Another objective of tomato research is the desire to produce a better-tasting "vine-ripe" tomato that has a greater shelf life and will resist bruising during shipment. Currently, supermarket tomatoes are picked green, then artificially ripened just before shipment. This process has ensured that the tomatoes resist bruising, but flavor is affected. Biotechnology has been used to develop the Flavr Savr tomato, which resists spoilage longer than conventional tomatoes.

Production of BD foods

Current techniques of developing organisms used in the production of BD foods typically involve the transfer to the host of the desired gene or genes in combination with a promoter and a gene for a selectable marker trait that allows the efficient isolation of cells or organisms that have been transformed from those that have not. Common selectable markers in plants have included resistance to antibiotics or herbicides. Several key issues have been raised with respect to the potential toxicity associated with BD foods, including the inherent toxicity of the transgenes and their products, and unintended effects resulting from the insertion of the new genetic material into the host genome. Unintended effects of gene insertion might include an over-expression by the host of inherently toxic or pharmacologically-active substances, silencing of normal host genes, or alterations in host metabolic pathways. It is important to recognise that, with the exception of the introduction of marker genes, the process of genetic engineering does not, in itself, create new types of risk. Most of the hazards listed above are also inherent in conventional breeding methods.

The guiding principle in the evaluation of BD foods by regulatory agencies in Europe and the USA is that their human and environmental safety is most effectively considered relative to comparable products and processes currently in use. From this arises the concept of "substantial equivalence." If a new food is found to be substantially equivalent in composition and nutritional characteristics to an existing food, it can be regarded as being as safe as the conventional food and does not require extensive safety testing. Evaluation of substantial equivalence includes consideration of the characteristics of the transgene and its likely effects within the host, and measurements of protein, fat and starch content, amino acid composition and vitamin and mineral equivalency together with levels of known allergens and other potentially toxic components. BD foods can either be substantially equivalent to an existing counterpart, substantially equivalent except for certain defined differences, or be non-equivalent, which would mean that more extensive safety testing might be necessary.

The examination of substantial equivalence therefore may only be the starting point of the safety assessment. It provides a valuable guide to the definition of potential hazards from BD foods and illuminates necessary areas for further study. While there is some concern relative to the meaning of "substantial" and how equivalency should be established, and debate over its use continues, the concept appears to be logical and robust in assessing the safety of foods derived from both genetically-modified plants and microorganisms. If it can be established with reasonable certainty that a BD food is no less safe than its conventional counterpoint, it provides a standard likely to be satisfactorily protective of public health. It is also an approach that has the flexibility to evolve in concert with the field of transgenic technology.

Humans typically consume a minimum of 0.1 to 1 gram of DNA in their diet each day. Therefore, the transgene in a genetically engineered plant is not a new type of material to our digestive systems and it is present in extremely small amounts. In transgenic corn, the transgenes represent about 0.0001% of the total DNA. Decades of research indicate that dietary DNA has no direct toxicity itself. On the contrary, exogenous nucleotides have been shown to play important beneficial roles in gut function and the immune system. Likewise, there is no compelling evidence for the incorporation and expression of plant-derived DNA, whether a transgene or not, into the genomes of consuming organisms.

Defense processes have evolved, including extensive hydrolytic breakdown of the DNA during digestion, excision of integrated foreign DNA from the host genome, and silencing of foreign gene expression by targeted DNA methylation that prevent the incorporation or expression of foreign DNA. Although much remains to be learned about the fate of dietary DNA in mammalian systems, the possibility of adverse effects arising from the presence of transgenic DNA in foods either by direct toxicity or gene transfer is minimal. The potential toxicity of the transgene product must be considered on a case-by-case basis. Particular attention must be paid if the transgene produces a known toxin (such as the Bacillus thuringiensis (Bt) endotoxins) or a protein with allergenic properties.

The level of risk of these gene products to consumers and those involved in food production can be and is evaluated by standard toxicological methods. The toxicology testing for the Bt endotoxins typifies this approach and has been described in detail by USEPA. The safety of most Bt toxins is assured by their easy digestibility as well as by their lack of intrinsic activity in mammalian systems. In this case, the good understanding of the mechanism of action of Bt toxins, and the selective nature of their biochemical effects on insect systems, increases the degree of certainty of the safety evaluations. However, each new transgenic product must be considered individually based on exposure levels and its potency in causing any toxic effects, as is typical of current risk assessment paradigms for chemical agents.

Allergenicity is one of the major concerns about food derived from transgenic crops. However, it is important to keep in mind that eating conventional food is not risk-free; allergies occur with many known and even new conventional foods. For example, the kiwi fruit was introduced into the U.S. and the European market in the 1960' s with no known human allergies; however, today there are people allergic to this fruit. The issues that have to be addressed regarding the potential allergenicity of BD foods are:

- do the products of novel genes have the ability to elicit allergic reactions in individuals who are already sensitised to the same, or a structurally similar, protein?
- will transgenic techniques alter the level of expression of existing protein allergens in the host crop plant?
- do the products of novel genes engineered into food plants have the ability to induce de novo sensitisation among susceptible individuals?

Considerable scientific resources are being committed to determine the most appropriate and accurate approaches for identifying and characterising potentially allergenic proteins. The first systematic approach to allergenicity assessment was developed by the International Life Sciences Institute (ILSI) in collaboration with the International Food Biotechnology Council and published in 1996. The hierarchical approach described therein has been reviewed and revised by the World Health Organisation (WHO) and the Food and Agriculture Organisation of the United Nations (FAO). The main approaches currently used in the evaluation of allergenicity are:

1. *Determinations of structural similarity, sequence homology and serological identity:* The objective is to determine whether, and to what extent, the novel protein of

interest resembles other proteins that are known to cause allergy among human populations. There are essentially three generic approaches. The first is to examine the overall structural similarity between the protein of interest and known allergens. The second is to determine, using appropriate databases, whether the novel protein is similar to known allergens with respect to either overall amino acid homology, or with respect to discrete areas of the molecule where complete sequence identity with a known allergen may indicate the presence of shared epitopes. The third approach is to determine whether specific IgE antibodies in serum drawn from sensitised subjects are able to recognise the protein of interest.

2. *Assessment of proteolytic stability*: There exists a good, but incomplete, correlation between the resistance of proteins to proteolytic digestion and their allergenic potential, the theory being that relative resistance to digestion will facilitate induction of allergic responses provided the protein possesses allergenic properties. One approach therefore is to characterise the susceptibility of the protein of interest to digestion by pepsin or in a simulated gastric fluid. However, this approach alone may not be sufficient to identify cross-reactive proteins with the potential to elicit allergic responses in food- or latex-sensitised individuals as in the case of oral allergy syndrome or latex-fruit syndrome. Nor are considerations of stability to digestion necessarily relevant for allergens that act through dermal or inhalation exposure and that may have significance for worker health. In these cases, other approaches such as structural homology searches and the use of animal models may be effective in identifying potential new allergens.

3. *Use of animal models:* Currently there are available no widely accepted or thoroughly evaluated animal models for the identification of protein allergens. Nevertheless, progress is being made and methods based on the characterisation of allergic responses or allergic reactions in rodents and other species have been described.

Although testing strategies for allergens are still evolving and no single test is fully predictive of human responses, when used in combination, allow scientists to address questions of potential allergenicity and these will increase in precision and certainty with time. Considerations of this type led the US federal agencies to deny approval of StarLink corn for human consumption because of the possibility that its Bt protein, Cry9C, may be a human allergen. This protein had been modified to slow its digestion and prolong its effect in the insect gut and this change rendered the protein less digestible in the human gut as well. After the accidental introduction of StarLink corn into the human food chain, a limited number of illnesses among consumers were reported. These were investigated by the Centers for Disease Control who found no evidence that the corn products were responsible. The only documented case where a human allergen was introduced into a food component by genetic engineering occurred when attempts were made to improve the nutritional quality of soybeans using a Brazil nut protein, the methionine-rich 2S albumin. Allergies to the Brazil nut have been documented, and while still in pre-commercial development, testing of these new soybeans for allergenicity was conducted in university and industrial laboratories. It was found that serum from people allergic to Brazil nuts also reacted to the new soybean. Once this was discovered, further development of the new soybean variety was halted and it was never marketed.

Concern has been expressed about the randomness with which genes are inserted into the host by current genetic engineering processes. This could, and does, result in pleiotropic and insertional mutagenic effects. The former term refers to the situation where a single gene causes multiple changes in the host phenotype and the latter to the situation where the insertion of the new gene induces changes in the expression of other genes. Such changes due to random insertion might cause the silencing of genes, changes in their level of expression or, potentially, the turning on of existing genes that were not previously being expressed. Pleiotropic effects could be manifested as unexpected new metabolic reactions arising from the activity of the inserted gene product on existing substrates or as changes in flow rates through normal metabolic pathways.

Unexpected and potentially undesirable pleiotropic or mutagenic changes in the genome of the host do occur, but these would likely be revealed by their effects on the development, growth or fertility of the host, or by the extensive testing of its chemical composition compared with isogenic untransformed plants which is a necessary part of any safety evaluation of transgenic crops. In the US, since 1987, the USDA Animal and Plant Health Inspection Service has completed over 5000 field trials with more than 70 different transgenic plant species. The only unexpected result was a mutation in a color gene and gene silencing through changes in the methylation status of these genes that led to unexpected color patterns in petunia flowers. Both of these effects are also seen in conventional plant breeding. While the possibility of an undetected increase in a toxic component in a new food cannot be entirely eliminated, the current safeguards make this unlikely and no toxicologically or nutritionally significant changes of this type are evident in the transgenic plants so far marketed for food production.

Substantial public concern about the safety of BD products was raised in 1989 when a number of cases of eosinophilia-myalgia syndrome (EMS) were reported among users of the amino acid tryptophan as a dietary supplement. By mid-1993, 37 deaths had been attributed to this outbreak. The development of the syndrome appeared among users of some batches of the supplement after a change in the manufacturing process that included the use of a new genetically modified microorganism in the fermentation. However, concomitant with this change were additional alterations in certain filtration and purification steps used previously in the manufacturing process. The exact cause of the outbreak and the nature of the toxic impurity have not been established with certainty. Thus, it is not possible to determine whether the change in purification, the genetic engineering of the organism, or some other factor or factors were to blame. A subsequent investigation revealed that cases of EMS occurred among consumers of tryptophan before the GM organism was introduced into the manufacturing process, although at a lower incidence.

The genetic modifications might have caused an increase in the level of the agent which was responsible for tryptophan-associated EMS, but it did not create a novel toxicant. This event is troubling in that the tryptophan would be regarded as highly purified and no adequate animal model has been found to replicate EMS, a probable autoimmune disease. This illustrates that toxicology has limits in its ability to explain and predict adverse effects in humans. These examples indicate that careful analysis of the changes in BD organisms is necessary to ensure against unexpected alterations in the levels of toxins, allergens and essential nutrients. This will be particularly critical if, as seems likely, engineering of the synthetic pathways of secondary metabolites is undertaken in plants e.g. to increase their resistance to insects and pathogens or produce compounds of pharmaceutical value. Such changes might create new and unanticipated secondary compounds with unknown toxic properties. New approaches to profiling changes in metabolites, proteins, and gene expression may be helpful in such cases.

The development of antibiotic resistance among pathogenic bacteria is a significant human health issue. However, no contribution to antibiotic resistance in gut bacteria arising from antibiotic resistance markers in BD foods has been documented. For several reasons, including the efficient destruction of the resistance gene in the human gut and the very low intrinsic rate of plant-microbe gene transfer, any contribution from this source is expected to be extremely small. Genes for resistance to kanamycin and related antibiotics already occur quite commonly in the environment including the flora of the human gut which naturally contains about 1 trillion kanamycin- or neomycin-resistant bacteria. Even if the occasional transfer of resistance from plant to bacterium did occur, the practical impact would be negligible. However, since any increase in antibiotic resistance is recognised as undesirable, and the technology is now available to omit the use of such marker genes, future genetically-modified organisms are unlikely to contain them. Thus concerns related to their use are likely to diminish.

In the USA, the FDA is entrusted with assuring that the nutritional composition of BD foods is substantially equivalent to that of the non-modified food. Studies are performed to determine whether nutrients, vitamins and minerals in the new food occur at the same level as in the conventionally-bred food sources. A typical example is the case of Roundup Ready soybeans. In this case, the protein, oil, fiber, ash, carbohydrates and moisture content and the amino acid and fatty acid composition in seeds and toasted soybean meal were compared with conventional soybeans. Fatty acid compositions and protein or amino acid levels of soybean oil were compared and special attention was given to checking the levels of antinutrients typically found in soybeans, e.g., trypsin inhibitors, lectins and isoflavones.

One difference between the conventional and non-conventional soybeans was detected in defatted, non-toasted soybean meal, the starting material for commercially utilised soybean protein which is not itself consumed. In this material, trypsin inhibitor levels were 11-26% higher in the transgenic soybeans. The levels of the trypsin inhibitors were similar in all lines in the seeds and in defatted, toasted soybean meal, the form used in foods. Except for this difference in trypsin inhibitor levels, all other nutritional aspects were equivalent between the transgenic line and the conventional soybean cultivars. Feeding studies demonstrated that there were no evident differences in nutritional value between the conventional and transgenic soybeans in rats, chickens, catfish and dairy cattle. Domestic animal feeding studies with a number of other transgenic crops have similarly shown no significant adverse changes in nutritional value.

In addition to the general concerns addressed that relate to food safety, additional attention is needed when the gene product is pesticidal or otherwise may be toxic to non-target organisms that consume it. The effects of each transgene product that is designed for pesticidal effects must be evaluated on a case-by-case basis against target and non-target organisms under specific field growth conditions for each transgenic crop. The foremost current example of this is the incorporation of Bt genes into crop plants for insect control. The toxic properties of Bt endotoxins to both target and non-target species of many kinds are well known. They show a narrow range of toxicity limited to specific groups of insects, primarily Lepidoptera, Coleoptera or Diptera, depending on the Bt strain. Nevertheless, Bt-producing plants have been tested broadly to determine whether any alteration in this limited spectrum of toxicity has occurred, without the discovery of any unexpected results. Exotoxins and enterotoxins, which are much more broadly toxic than the endotoxins, are also produced by some Bt strains, but these are not present in the transformed plant because their genes are not transferred into the crop.

In plants transformed with Bt genes to control lepidopterans, toxicity to non-target lepidopterans would be expected if exposure occurs by feeding on the transformed crop. Particular concern has been expressed over the potential toxicity of the Bt toxin in corn pollen to the Monarch butterfly after initial laboratory studies showed increased mortality in larvae fed on leaves dusted with transgenic pollen. However, most transgenic corn pollen contains much lower, non-lethal levels of Bt toxins than the strain used and there is only a limited synchrony between the feeding period of the most sensitive younger larvae and the period when corn pollen is shed. Also, corn pollen does not typically move far beyond the borders of the field, leaving significant amounts of milkweed uncontaminated in many locations.

It is unlikely that a substantial risk to these butterflies exists in the field since only a negligible portion of the population is exposed to toxic levels of Bt. Beyond the question of the potential toxicity of Bt corn to such valued insects, it is also important to recollect that the common alternative is to spray corn with synthetic insecticides, which are not as selective as the Bt toxin. In a sweet corn field containing milkweed plants and treated with a synthetic pyrethroid for insect control, 91-100% of the monarch butterfly larvae placed on the milkweed leaves after spraying were killed. In plots where Bt sweet corn was planted and the pollen fell naturally on the milkweed leaves, larval death rates were much lower (7-20%) and indistinguishable from those in untreated non-Bt corn plots.

Antibiotic Resistance Markers

Antibiotic resistance genes have the ability to selectively inactivate certain antibiotics and consequently protect cells against these antibiotics. An antibiotic resistance gene can thus be used to "tag" a gene carrying a trait or characteristic of interest. In practice the antibiotic gene is linked to a gene carrying the trait of interest prior to its introduction into the recipient cells, whether of bacterial, yeast, plant or animal origin. These cells are then incubated in the presence of the antibiotic. The only cells which multiply under these conditions are the ones that have incorporated the antibiotic resistance gene together with the trait of interest. The identification of the "transgenic cells" would be extremely tedious, and even impossible, without this selection procedure since only a very small fraction of the cell population incorporates the introduced genes. A selection procedure is a necessity in genetic engineering, and this is the reason why antibiotic resistance genes have been so widely use in many different areas of biotechnology for a number of years.

Recent approvals for the commercialisation of Genetically Modified (GM) crops containing antibiotic resistance markers have raised concern in Europe about the risk of spreading antibiotic resistance genes to previously unsusceptible microbes and hence making them resistant to the antibiotics used. Public concern about this topic is widespread and the issue is repeatedly raised in the media. As a consequence governments in the European

Union have recommended "phasing out" GM crops containing antibiotic resistance markers contrary to the recommendations of several national and European scientific committees.

Bacteria are microbes that are present everywhere in the environment and in plants and animals. Microbes occupying the same habitat compete for nutrients and for their own survival some have evolved naturally to produce antibiotics to eliminate their competitors. Antibiotics inhibit a cell' s growth by blocking some of its essential metabolic processes. Bacterial strains producing a given antibiotic therefore have to carry resistance to inactivate the corresponding antibiotic and thus prevent its own self-destruction. In the evolutionary race between microbes the production of new antibiotics is usually countered through the development of resistance mechanisms both by producing- and target-organism. There is in nature a wide range of antibiotic and corresponding antibiotic genes. However, rather than developing their own resistance mechanisms, targeted bacteria will in general acquire antibiotic resistance genes which are already present in the bacterial pool surrounding them. This is facilitated due to the fact that bacteria are generally quite promiscuous about exchanging genetic material between each other.

The presence of an antibiotic confers an advantage to a resistant bacterium and so under these conditions resistance development and spread increases. The discovery of antibiotic action against disease-causing agents at beginning of the 20th century had a major impact on medicine. However, it has been used without sufficient limitation. Man' s use of antibiotics has drastically increased their global distribution and consequently promoted the spread of resistant microbes. Increased use of antibiotics in clinical and also veterinary medicine is the major cause of the increasing incidence of antibiotic resistance in bacteria. In addition, antibiotics have been and still are used extensively as animal feed additives which has brought about the selection of resistant bacteria in healthy animals. Antibiotics are also widely sprayed on crop plants, orchards, vines, etc to protect against pathogens.

Today resistance to antibiotics is so widespread that some of the first generation of antibiotics are of no use anymore. Multiple antibiotic resistance in pathogenic strains of Staphylococcus and Mycobacterium tuberculosis, especially in hospitals, is of particular concern to the medical community. Concern about the use of antibiotic resistance markers in relation to overall antibiotic use management is expressed. Since the mid-1980s modern biotechnology methods have been developed to improve agricultural crops through the introduction of genetic material that confers advantageous characteristics. There are two types of antibiotic resistant marker (ARM) genes used in transgenic plants:

1. *Genes driven by bacterial promoters.* These genes have been used during the initial stages of the assembly of the pieces of DNA intended for transfer into the plant cells. The purpose of these genes was to select for the amplification of the pieces of constructed DNA in the receiving bacteria. The gene providing resistance to ampicillin belongs to this category. Genetically modified plants containing those genes are from the earliest generation of technologies and present technologies only allow the removal of these genes before initiating the plant transformation process.

2. *Genes which allow the selection of plant cells which have up taken the piece of DNA carrying the trait or characteristic of interest.* The insertion of a gene into a plant cell by transformation is a very inefficient

process since only a few thousand cells of the many millions used take up the desired gene. The transfer of an antibiotic resistance marker gene together with the gene of interest allows these very few cells to be selected as only those cells that have taken up both genes will survive and multiply in the presence of the corresponding antibiotic in the growth medium. A genetically modified plant is then grown from these modified cells and the marker is no longer needed.

The time needed for the development of a genetically modified crop with a new trait usually exceeds 10 years. Safety aspects are taken into consideration at each step of product development. This starts at the selection of appropriate proteins and genes to insert into the plant, followed by experiments designed to assess potential impacts on human health from the consumption of the GM crop and then several years of field trials with the genetically modified crop to assess potential environmental safety impacts. Antibiotic resistance markers used in the development of genetically modified crops have been selected by scientists according to various safety criteria. These include that the marker genes occur frequently in natural microbial populations and that they confer resistance to a narrow range of specific antibiotics with limited application in human and veterinarian medicine.

The most widely used antibiotic resistance marker for the selection of transformed plant cells is the *nptII* gene, also called *aph(3')-II,* which confers resistance to the antibiotics neomycin and kanamycin. This gene is present in ten of the fourteen GM plants containing antibiotic resistant marker gene submitted for marketing in the EU. For example, it has been used to develop the delayed ripening tomato, herbicide-tolerant and insect-protected corn and cotton varieties. The choice of using this antibiotic resistant marker gene has been driven by the fact that the antibiotics kanamycin and neomycin are not important in medical treatment and that, on average, 20 to 40% of the bacteria that occur naturally in human or animal digestive tracts are already resistant to kanamycin. Kanamycin/neomycin resistant bacteria are ubiquitous in nature, their prevalence being dependent upon the source of the bacteria isolated, the highest level being found in pig manure.

SAFETY ISSUES

The safety assessment of crops containing antibiotic resistance markers has been thoroughly reviewed by experts from internationally recognised scientific bodies and scientific committees in the European Union and governmental experts from Canada, Japan, Switzerland and the United States among other countries. The experts evaluated the safety of crops containing the *nptII* marker gene conferring resistance to the antibiotics kanamycin and neomycin, the aad gene conferring resistance to streptomycin and spectinomycin and the bla gene conferring resistance to ampicillin. They concluded that the risk of gene transfer from transgenic crops containing these genes to the microbial population is negligible and, if it were to occur, there would be no impact on the present high frequency of occurrence of antibiotic resistance in microbial populations. The European Commission Scientific Committee on Plants provided an opinion on the safety of a Bt insectprotected maize product containing the *nptII* gene. Their conclusions are based on the following observations:

1. The antibiotic resistance markers used in plant biotechnology were obtained from naturally occurring bacteria. The *nptII* gene in particular was isolated from bacteria in the human gut. The markers in approved products are already widespread because effective

natural transfer mechanisms exist to shuttle the genes directly between bacterial cells. This process confers advantages to bacteria. Some microbial populations can hence act as reservoirs for certain antibiotic resistance genes that then are rapidly spread in response to selective pressures.

2. The antibiotics inactivated by *nptII*, kanamycin and neomycin are rarely used in human therapy since they have been replaced by less toxic antibiotics with greater efficacy.

3. There is no known mechanism for the transfer of genes from plant cells to bacteria. Gene transfer to soil micro-organisms could in theory occur in fields where GM crops are cultivated. However, such events have not yet been detected. Both bacterial and plant DNA have been shown to persist in soil for weeks or months by binding to the surfaces of soil particles. The availability over time of plant DNA, an important factor in determining likelihood of gene transfer, is therefore likely to be higher in the soil than in the gut. However, while certain bacteria have the ability to take up naked DNA spontaneously, numerous barriers would need to be overcome before intact genes from a plant could be taken up from a natural environment, and be functional and maintained in bacteria; such an event has never been shown to occur under natural conditions. Even if transfer were to occur at a very low frequency, this would not add significantly to the number of antibiotic resistant bacteria that already exist in nature.

4 The rate of DNA transfer between bacteria under optimal conditions in nature is in the range of 10-2 to 10-5. The estimated frequency of uptake of the *nptII* gene marker from transgenic plants to bacteria under

optimal conditions is 10-17 and thus considered to be insignificantly small. The *nptII* gene corresponds to only 0.00004% of the total maize genome and would compete with the rest of the DNA for uptake by the bacterium. The availability of free GM plant-derived DNA in the rumen or the gastro-intestinal tract is further limited due to rapid digestion by pancreatic fluids and acid saliva. The participation of *nptII* gene from transgenic plants to the overall pool of kanamycin resistant bacteria appears even more insignificant when compared to the large pool of genes providing resistance towards kanamycin that already exists in bacteria and the high level of gene transfer that is constantly occurring in bacterial populations. Therefore a bacterium is several orders of magnitude more likely to acquire a resistance gene from another bacterium rather than from the DNA of the GM crop.

5. Antibiotic resistance markers only confer resistance against specific antibiotics. Antibiotic resistance markers do not result in antibiotic production. There are therefore no antibiotics present in food produced from plants produced using biotechnology.

Alternative selectable markers for plants fall into two categories. Some markers confer resistance to chemicals other than antibiotics that kill plant cells such as herbicides and lethal concentrations of the amino acids lysine and threonine. The enzyme that confers resistance to high concentrations of lysine and threonine can interfere with amino acid biosynthesis and if expressed at high levels cause abnormal plant development. The relevant genes are therefore not suitable as marker systems. The presence of herbicide-tolerance markers may be undesirable. Glyphosate is the most effective herbicide to control potato seed from previous years in temperate regions with mild winters and therefore glyphosate-tolerant potatoes would not be suitable in these regions. Furthermore, plants containing herbicidetolerance marker genes which are not registered as herbicide-tolerant crops may tempt the misuse of herbicides.

Other alternative marker systems rely on the growth of plant cells in the presence of unusual nutrients, including cytokinin, glucuronides, xylose or mannose, which will not allow nontransformed plant cells to grow. For example, plant cells usually do not use mannose as a source of sugar. The delivery of a gene allowing mannose to be metabolised in plant cells and the subsequent cultivation of those cells in a medium containing mannose as the sole source of sugar would allow only those cells which have taken up the gene to grow. When these systems, which are still in their development phase, work reliably on a large scale in a wide range of different environments risk assessments will have to be conducted to assess the potential ecological impacts of plants that can grow on a new substrate, the impact on the overall plant metabolism and the consequences on human or animal diet from increased levels of metabolites in these crops that might not be present in the conventional counterpart.

It is not possible to remove marker genes once they are integrated into a plant genome unless a particular mechanism for removal is incorporated along with the marker gene and the gene of interest at the time of the transformation. As was mentioned above, it is possible to avoid introducing into plant cells antibiotic resistant marker genes which are only used for the assembly and amplification of the DNA constructs in bacteria, and therefore are not necessary during the plant step of the transformation procedure. The removal prior to commercialisation of marker genes which are driven by plant promoters and are used for selection of plant cells has become the aim of both consumers and industry.

Extensive research with this aim is being carried out both by industry and academic institutions. Among the technologies being assessed are:

1. *The use of meganucleases.* These are enzymes which can specifically recognise long DNA sequences. These recognition sequences are introduced on both sides of the antibiotic resistant marker gene to be introduced into the plant cell. Once the transformed cells have been selected on the corresponding antibiotic, the meganuclease is introduced into the plant cell, and will allow the excision of the antibiotic resistant marker gene. This technology has proven to be very efficient in certain plants, but difficult to handle in others possibly because the meganuclease recognises sites in the plant genome itself.

2. The presence of homologous DNA sequences on both side of the antibiotic resistant marker gene may allow for random recombination and elimination of the gene. This process of homologous recombination occurs at low frequency and may be plant specific.

3 *It is possible to introduce the trait of interest and the antibiotic resistant marker on different DNA constructs.* Following transformation, each molecule integrates on a different chromosome. In this case it is possible to segregate the trait of interest from the marker gene at the next generation. Frequencies of integration on separate chromosomes can be quite low when compared to integration at the same locus. Factors involved in regulating these frequencies are not yet understood.

Extensive effort is being invested by researchers to develop these technologies. Some products using one or other of these strategies are already in the pipeline. But today none of these approaches can routinely be used for every crop.

The development of these technologies will probably be limited in the first place to laboratories having a strong infrastructure. These laboratories will have to be able to afford to develop a greater number of transgenic plant lines containing the antibiotic marker gene than will be needed for the subsequent screening for the elimination of the marker gene.

The concern about antibiotic resistance marker genes is predominantly about their transfer and expression in bacterial cells and a technology which would prevent such expression might have to be considered. Already the genes used for selection in plants are controlled by plant promoter sequences which render them unlikely to be sufficiently expressed in bacteria. The introduction of an intron sequence in the marker gene would restrict its expression to plant cells and definitively prevent any expression in bacteria. Introns are sequences of DNA that naturally interrupt the coding sequences of animal and plant cells. These are equipped with mechanisms allowing their removal during the transcription process while bacteria are not equipped to do this and therefore would be unable to read a gene containing introns.

Antibiotic resistance marker genes such as the *nptII* gene which provides resistance towards kanamycin are very well researched in all the relevant aspects such as their functioning, biochemical properties and prevalence in the bacterial community. Their safety has been well examined and assessed. Under these conditions it is likely that achieving the same level of confidence as has been established for *nptII* with another selection system may be long and difficult. Any remaining concerns attached to such genes could be removed by the addition of an intron.

Scientific experts agree that the main cause for the spread of antibiotic resistance is the overuse of antibiotics in human and veterinarian medicine. Public concerns on the use of GM crops with antibiotic resistance markers are, however, widespread, especially since this issue was repeatedly raised in the media. Several governments in the European Union have recommended the phasing out GM crops containing any antibiotic resistance markers. However, the risk of compromising the efficacy of antibiotics is considered to be vanishingly small. Most alternatives are still in their development phase, are not widely available and will be difficult to implement in a less developed country.

BIOTECHNOLOGY AND RURAL PEOPLE

The use of biotechnology in agriculture and food production could elicit concerns similar to those expressed about agricultural chemicals. In addition, other dimensions of biotechnology are also starting to draw public attention. Researchers at North Carolina State University conducted telephone interviews with 1,228 adults from across the country during early 1992. Results showed that most people were in favor of many of the potential applications. Two-thirds of the people surveyed felt that biotechnology would personally benefit people like themselves within the next five years. Just as many disagreed with the statement that only the companies that make the products will benefit from biotechnology. Three-quarters of the people surveyed felt that biotechnology will have a positive effect on food quality and nutrition. About two-thirds felt that biotechnology will also have a positive effect on environmental quality.

Consumers found certain applications to be more acceptable than others. Insect-resistant and herbicide-resistant plants were the most acceptable types of applications. Consumers were less accepting of animal related products or food additives. Survey results indicate that for many consumers lower prices will be more important than improved quality as an incentive for consumption of biotechnology products. Overall, two-thirds of the respondents supported the use of biotechnology in agriculture and food production. Certain groups were more likely to support the use of biotechnology and expressed greater willingness to buy the products. Men were more likely to accept the use of biotechnology, as were respondents with higher levels of education.

Biotechnology was less acceptable to people who had not heard or read much about it or who expressed lower interest. The survey did reveal some consumer concerns about the use of biotechnology in agriculture and food production. People appeared worried about the potential impacts of biotechnology on the natural environment. Moral and ethical concerns also appear relevant. About one-quarter of the respondents felt it is morally wrong to use biotechnology to change plants, and over one-half felt that use of biotechnology to change animals is morally wrong. The process of biotechnology will itself be of some importance to consumers and that the source of the genetic material used will be relevant to them. For example, about two-thirds of the respondents said that more nutritious potatoes would be acceptable if genes were added from another plant such as corn. However, only about one-quarter would find such potatoes acceptable if the genes were added from an animal.

People may believe that transferring genes among different organisms involves more than the simple movement of DNA that carries a specific desirable trait. Over one-half of the respondents had read or heard little or nothing about biotechnology or its specific applications. However, the results also show a good deal of consumer interest in learning more about biotechnology. About two-thirds of all respondents expressed some or a lot of interest in learning more about biotechnology.

Most of the public's views on this subject will be shaped by the media information they receive, as well as by discussions with family and friends. The survey clearly points out some educational needs and challenges. It is important to put biotechnology into a historical and technical context related to other types of food production technologies. The survey indicated limited understanding of traditional agricultural practices such as cross-breeding. It will be necessary, therefore, to show how biotechnology represents an incremental change rather than a drastic departure from past practices. The survey also indicates the need to more clearly differentiate the use of biotechnology from the use of pesticides and other chemical technologies. Some confusion may exist along these lines.

Public participation in the regulatory process and confidence in such regulations will be vital to ensure greater confidence in the ability of the government to manage genetic engineering effectively. The survey indicated a fairly low level of confidence in the ability of government agencies to regulate biotechnology effectively. Low confidence was also reflected in the fact that almost all respondents felt that government should pay more attention to what people like themselves think about biotechnology. In fact, over 80 percent agreed that citizens have too little say in decisions about whether or not biotechnology should be used. People clearly want more voice in decisions that affect their lives. This situation is not unique to biotechnology; it includes other areas of science and technology as well. An informed public will be an important partner in public policy decisions.

At the time of the survey, consumers appeared cautiously optimistic about the use of biotechnology in agriculture and food production. However, it also appears that consumer acceptance of food produced through biotechnology could be limited by one or more of the following factors: insufficient knowledge; lack of confidence in government; perceived exclusion from decision-making processes; moral objections to tampering with nature; and perceived environmental and food safety concerns. It will also be important to realize that not all products of biotechnology will be evaluated equally. Acceptance of biotechnology appears to vary depending on the nature of the application, including the perceived benefits and risks of each. Further education about the potential benefits and risks of the various products of biotechnology will be vital.

The Society of Toxicology (SOT) is committed to protecting and enhancing human, animal and environmental health through the sound application of the fundamental principles of the science of toxicology. It is with this goal in mind that the SOT defines here its current consensus position on the safety of foods produced through biotechnology. In this context, biotechnology is taken to mean those processes whereby genes that are not endogenous to the organism are transferred to microorganisms, plants or animals employed in food production, or where the expression of existing genes is permanently modified, using the techniques of genetic engineering.

Conventional techniques of plant and animal breeding, which are not considered here, also involve genetic modification. The extent of the genetic changes resulting from such conventional breeding techniques, which is generally undefined, far exceeds that typically produced by transgenic methods. Consequently, it is important to recognise that it is the product, and not the process of modification, that is the focus of concern regarding the human or environmental safety of biotechnology-derived (BD) foods.

The principal responsibilities of toxicologists are to define and characterise the potential for natural and manufactured materials to cause adverse health effects and to assess, as accurately as possible, the plausibility and level of risk for human or animal health or for environmental damage under a defined set of circumstances. It is not the task of the Society of Toxicology to determine the overall value of a product or process by balancing health or environmental risks with potential benefits, or to choose between different strategies to manage risk, although toxicological considerations are important in both processes. Major areas of concern in the development and application of such foods in agriculture relate to the possibility of deleterious effects on both human health and the environment.

3
GMC in Biotechnology

Genetic engineering could improve the growth rate and yield from crops, and so feed more people from the available land, in the face of the expansion of the world's population. This is not as simple as might be expected, however. It works quite well for some species of plants but not others. It is hardly successful at all in farm animals, where it seems that the main application of the genetic revolution will be in identifying genetic "markers" to assist normal selective breeding, or in novel applications like producing pharmaceuticals in sheep's milk. In theory it may be possible to develop genetically modified crops (GMC) to be more resistant to the vagaries of climate, or to grow on marginal land prone to drought or erosion, or highly salt or acid conditions. Potentially, this could greatly assist food production in vulnerable areas of the world where agriculture is on the margins of viability. Unfortunately this is rather difficult technically. Although there are some Third World examples, to date most of the research in this area seems to be directed not to developing country situations but to western crop agriculture.

The most successful applications so far have been to engineer plants to be genetically resistant to pests, diseases or weeds. This aim is to require smaller quantities and less virulent chemicals, in a way difficult to achieve by ordinary plant breeding. One approach is to devise a herbicide and genetically modify crop plants to be resistant to it, so that when the herbicide is sprayed on to a field it kills all the weeds but leaves the crop intact. Genetic engineering is not always as straightforward and specific as may be claimed. The other main approach is to introduce genes from other plants, micro-organisms or even the pest itself.

Genetic Engineering Techniques for Disease Resistance

A wide range of diseases and pests, including viruses, fungi, bacteria, insects, mites and plant nematodes are involved in significant crop losses each year. Today, the damage from insects has become more severe. Farmers have been trying to minimise losses from crop pests for hundreds of years. In the past, they have used conventional breeding practices, such as hybridisation, to develop crops with better pest resistance, or chemical insecticides or biological control systems, such as predator insects that attack the targeted crop pests.

Scientists can now make plants more pest resistant by inserting specific resistance genes from other plants or organisms. In some cases, recombinant DNA techniques were the first methods used to do this. This first generation of genetic engineering techniques for disease resistance relied mainly on affecting single gene traits. However, many resistance traits, such as those for fungal resistance, involve the interaction of several genes. Thus, future genetic engineering strategies are concentrating on means to control multiple gene transfers.

Efforts to develop resistance to several pests or pathogens will require the use of many gene transformations. The development of pest resistance in plants remains an ongoing effort, however, as pests themselves acquire new invasive strategies and become resistant to control measures. While development of pest-resistant plants is underway in public and private laboratories, the time required to create resistant strains, breed them into stable varieties, perform field-testing and obtain regulatory approval has so far limited the number of genetically engineered varieties commercialised.

Efforts to build crops resistant to diseases are occurring on multiple fronts. One approach is to find varieties of plants that demonstrate resistance to a specific infection-causing organism, and then determine the genetic components responsible for this natural resistance. The responsible genes can then be transferred to plants that don't have them. Other strategies rely on identifying the genes within a plant responsible for generating substances that fight pathogens, and then learning how to enhance the plant's ability to make them.

Developing Virus Resistance Food crops

Viruses are among the most ubiquitous pests in agriculture. Scientists are working to develop viral resistance in a variety of crops including squash, potato, sweet potato, wheat, papaya and raspberries. Viruses are studied widely because they not only cause disease in humans, plants, animals and insects, but also are used as tools in the study of molecular biology and, in some cases, in the development of vaccines to fight the diseases they can cause. Several techniques for virus resistance have been developed. These include viral coat protein technology and multiple gene transfers. A viral coat protein acts like a vaccine, causing the plant to develop resistance to the particular virus. Transferring the gene for a viral coat protein, a part of the outer shell of a virus that does not cause disease, into a plant acts like a vaccine for the plant. The plant is then able to resist the virus, analogous to the way vaccines keep us from getting certain diseases like measles. The advantage of introducing only the coat protein is that it induces resistance without the introduction of the actual virus. The technique has been used successfully in many plants against several different viruses.

The first genetically engineered virus-resistant food crop in the marketplace was yellow crookneck squash. Using the viral coat protein approach, this squash was engineered to resist the watermelon mosaic virus and the zucchini yellow mosaic virus. Potatoes are highly susceptible to many viruses, including the potato mosaic virus and the potato leaf roll virus. A leaf roll virus epidemic in 1996 was responsible for heavy potato crop losses in Idaho. The virus, spread by aphids, damaged the potatoes to the point that they were unmarketable. Scientists in Mexico, in collaboration with researchers at Monsanto, have developed potatoes resistant to several forms of this virus. Research on disease-resistant potatoes is continuing at other laboratories. The feathery mottle virus has a damaging effect on sweet potatoes.

In 1991, researchers began genetically engineering varieties of sweet potato grown in Africa, where it is an important subsistence crop. The sweet potato was engineered with coat protein from this virus and replicase genes. Replicase is an enzyme involved in the duplication of certain viral RNA molecules. Current field-testing has demonstrated successful gene transformations and the desired development of resistance to sweet potato feathery mottle virus. Although wheat is an important food source, development of genetically engineered varieties has been slower than in corn, soy and cotton. A major pest in wheat is barley yellow dwarf virus, which can cause damage in major wheat-growing regions such as North Dakota, because no resistant strains are known. Work is in progress to engineer resistance to this disease using the viral coat protein technique. The wheat genome is highly complex-ten to twenty times larger than that of cotton or rice-and carries an exceptionally large amount of repetitive DNA sequences. Thus, targeting particular genes is challenging, and transgenic wheat biotechnology has advanced more slowly than that of other crops.

The papaya crop in Hawaii was nearly wiped out in the 1950s by the papaya ringspot virus (PRSV). Transmitted by aphids, this virus causes one of the most serious diseases of papaya worldwide. Work to develop a transgenic virus-resistant variety began in the late 1980s. By 1992, resistant lines were field-tested; approvals for commercialisation were granted in 1997. The transgenic-resistant papaya is now in wide use in Hawaii, and similar work is in progress in the Philippines, Malaysia, Thailand, Vietnam and Indonesia to enhance resistance in local papaya varieties where ringspot virus is a major pest. Researchers are also modifying other fruits for virus resistance.

Developing Fungi Resistance Food crops

The search for genetic engineering tactics to combat fungi has intensified with the need to find adequate substitutes for fungicides such as methyl bromide, widely used on fruit and vegetables but being phased out due to its links to ozone depletion. One emerging area is directed at a plant's production of defensins, a family of naturally occurring antimicrobial proteins which enhance the plant's tolerance to pathogens, especially bacteria. Certain defensins also demonstrate an ability to fight fungal infections. Defensins are found throughout nature in insects, mammals (including humans), crustaceans, fish and plants.

Defensins from moths and butterflies, the fruit fly, pea seeds and alfalfa seeds all show potent antifungal activity. The first transgenic application of defensins was the incorporation into potatoes of the antifungal defensin from alfalfa. Laboratory and field trials showed that the transgenic potatoes were as resistant to the fungal pathogen Verticillium dahliae as non-transgenic potatoes treated with fungicide. Although studies are continuing, the chance that fungi will build resistance to defensins is thought unlikely. No known resistant strains of bacteria or fungi have yet evolved that can overcome these highly protective, pesticidal proteins. Ongoing research involving banana and cassava is directed to cloning resistance genes for major tropical diseases such as black sigatoka, a leaf fungus that widely infects bananas, cassava mosaic disease and cassava bacterial blight.

In bananas, transgenic lines combining several antifungal genes have been generated. Selected lines are currently being tested for resistance to black sigatoka and Panama disease under greenhouse and field conditions. Scientists are devising protection against the plant fungus Botrytis cinerea, a serious pathogen in wheat and barley. The strategy uses the gene for a natural plant defence compound named resveratrol. Scientists have also introduced a gene from a wine grape into barley to confer resistance to Botrytis cinerea. Field trials are underway.

Resistance to potato late blight, a disease caused by Phytophthora infestans, receives high priority in potato research. Plant disease from this fungus can be destructive to crop production, as was dramatically illustrated in the Irish potato famine. In 1995, a U.S. late blight epidemic (caused by new aggressive strains of Phytophthora infestans) affected nearly 160,000 acres of potatoes, or about 20 percent of domestic production. Research is underway to genetically engineer potatoes that express the enzyme glucose oxidase and develop resistance to Phytophthora blights (Douches undated). At present, however, no products are close to commercialisation. Potatoes are also being transformed using a soybean gene for a protein (beta-1, 3-endoglucanase) that confers resistance to infection by Phytophthora.

Other studies report that transgenic potatoes expressing a protein called osmotin showed reduced damage from lesion growth in leaves inoculated with the Phytophthora infestans pathogen. Still other research is attempting to boost fungal resistance in potatoes by transferring resistance genes from peas. Infection of these transgenic potatoes with the fungus triggers hormone-like signals in the potatoes that turn on the pea resistance genes. One substance that is produced, chitosan, stops fungal growth and activates the potato's own natural defence systems. In rice, blast and sheath blight are major fungal diseases. Scientists created transgenic strains resistant to sheath blight that are currently being field-tested.

Developing Bacteria Resistance Food crops

Most food crops are susceptible to bacterial diseases, but bacteria rarely attack certain plants, such as mosses, ferns and conifers. Bacterial infections in plants may cause leaf and fruit spots (lesions), soft rots, yellowing, wilting, stunting, tumours, scabs or blossom blights. When tissue damage occurs on the blossoms, fruit or roots of food crops, yields may be reduced. Potatoes are susceptible to blackleg and soft rot diseases caused by the bacterial pathogen Erwinia carotovora.

To combat these bacteria, scientists have exploited the family of enzymes known as lysozymes that catalyze the breakdown of bacterial cell walls. Using cloned lysozyme genes and a promoter, transgenic potatoes were created that produced lysozyme. In laboratory tests, the transformed potatoes exhibited substantially enhanced resistance to Erwinia carotovora. Field tests and further development of resistant lines are in progress. A different transgenic strategy to combat Erwinia carotovora was demonstrated in tobacco engineered to overexpress a peptide that kills bacteria. The genetically engineered tobacco plants were resistant to both Erwinia carotovora and Pseudomonas syringae pv tabaci, the pathogen responsible for wild fire disease in rice. Scientists have also successfully transferred a bacterial resistance gene from wild rice to cultivated rice.

Developing Insects Resistance Food crops

There are several different combat tactics, including engineering for the expression of toxins in plants that kill insects when they consume the plant material, but are nontoxic to other species that eat the plant. Other alterations focus on inducing sterility in the pest organism or affecting the digestion or metabolism of the pests. In addition, attempts to enhance a plant's natural ability to produce leaf wax could make the plant more difficult for insects to consume. The best known and most widely used transgenic pest-protected crops are those that express insecticidal proteins derived from genes cloned from the soil bacterium Bacillus thuringiensis, more commonly known as Bt. Crystal (Cry) proteins or delta-endotoxins formed by this bacterium are toxic to many insect species. Delta-endotoxins bind specifically in the insect gut to receptor proteins, destroying cells and killing the insect in several days (shown below). There are several different Bt strains containing many different toxins. Scientists have identified and isolated the genes for several toxin proteins from different Bt strains.

In recent years, these genes have been introduced into several crop plants in an effort to protect them from insect attack and eliminate the need for spraying synthetic chemical pesticides. There are more than 100 patents for Bt Cry genes. Bt field corn, sweet corn, soy, potato and cotton are commercialised in the U.S., and one or more of these are commercialised in at least 11 other countries. Bt controls the larvae of butterflies and moths (Lepidopteran insects) that eat the plants. It is especially effective against the larvae of the European corn borer (shown left), a significant corn pest in the U.S., as well as the Southwestern corn borer and the lesser cornstalk borer. In sweet corn, Bt toxins effectively deter corn earworm and fall armyworm.

Recently, a different strain of Bt, Bacillus thuringiensis tenebrionis, was used as a gene source to confer resistance to corn rootworm, another major pest in cornfields. The resistant corn is currently in field trials. Bt hybrid rice is also undergoing field-testing and is showing considerable effectiveness in resisting major pests in Asia such as the leaf folder, yellow stem borer and striped stem borer. Bt canola is also under development.Borers also create a good environment for fungi to grow. Where fusarium fungi grow, they reduce plant quality and generate fumonisins-toxins that can be fatal to farm animals and have been linked to liver and esophageal cancer in African farmers. Thus, one way to reduce fungal contamination is to control pests.

Scientists have measured reductions in fumonisin levels in Bt corn of 90 percent or greater. Bt works against insects that eat plant tissue. However, those pests that do not eat the leaves, but rather pierce and suck nutrients from the plant, require different defence strategies. These insects include aphids, white flies and stink bugs. White flies are a major pest in poinsettias, sweet potatoes and cotton. Because these insects do not consume large amounts of plant material, a leading way to combat them is the genetic expression of toxic proteins that are strong enough to kill the pest, yet safe for the plant and non-target organisms.

Avidin in transgenic corn demonstrates a different approach. Avidin is a glycoprotein, an organic compound composed of both a protein and a carbohydrate, and is usually found in egg whites. Avidin is known for chemically tying up the vitamin biotin, making it unavailable as a nutrient. Insects eating transgenic corn modified to produce avidin die from biotin deficiency. Although this corn was not toxic to mice, further evaluation of its potential for insect toxicity and safety for human consumption is awaited. Plants produce wax as a natural protective coating. Genetic modification can increase the expression of this inherent trait. Experiments to increase leaf wax are in the early stages, but scientists have already raised wax content by as much as 15-fold. This strategy is aimed at increasing the plant's resistance to both pests and fungal pathogens.

Developing Nematode Resistance Food crops

The most common of nematode plant parasites found worldwide is the root-knot nematode. Probably every form of plant life, including field crops, ornamentals and trees, is attacked by at least one species of nematode. They are responsible for 10 percent of global crop losses worth an estimated $80 billion a year. Transgenic strategies to combat nematodes are emerging. Nematodes are particularly destructive in bananas, soybeans, rice and potatoes. Scientists are fighting these parasitic worms in potato and banana crops using the genes for cystatins, defence proteins that occur naturally in rice and sunflowers. Incorporation of the genes in potatoes produced as much as 70 percent nematode resistance in field trials.

Nematodes are particularly fond of soybeans. In the U.S., the soybean cyst nematode is considered the most devastating pest. Standard plant breeding led to a highly resistant variety of soybeans from a wild strain, but it did not cross well with modern soybean lines. Using genetic markers, a means of identifying cells with particular traits, scientists bred plants containing the resistance gene with domesticated varieties, circumventing the poor performance characteristics of the wild variety. While the new varieties are not transgenic, they resulted from combining the use of modern genetic markers with conventional breeding techniques.

Improving Field-crop Production and Soil Management

Improving field-crop production and soil management is another central aim of genetic engineering technology in commodity crops. Applications include crop resistance to herbicides; improved nitrogen utilisation, reducing need for fertiliser; increased tolerance to stresses such as drought and frost; regulation of plant hormones, which are key to plant growth and development; attempts to increase yield, and a multitude of other, less widespread applications.

There are many negative effects when weeds grow with crop plants, the most common being competition for sunlight, water, space and soil nutrients. If weeds grow with crops, they too use these growth factors, and may cause losses great enough to justify control measures. In addition to economic yield loss, other concerns may determine when weed control is justified. For example, eastern black nightshade in soybeans or late-emerging grasses in corn may not reduce yield, but these weeds can clog equipment, causing harvest delays. The most common method currently employed to manage weeds is the use of herbicides. The use of genetic modification techniques has created crops that are both tolerant and resistant to herbicides, or weed killers. This technology allows herbicides to be sprayed over resistant crops from emergence through flowering, thus making the applications more effective. To date, six categories of these crops have been engineered to be resistant to the herbicides glyphosate, glufosinate ammonium, imidazolinone, sulfonylurea, sethoxydim and bromoxynil.

Probably the best-known herbicide for which tolerance has been genetically engineered into crops is glyphosate, known commercially by brand names such as Roundup, Rodeo and Accord. Resistance to glyphosate is the transgenic trait most common in agriculture worldwide. To date soy, corn, cotton, canola, sugar beets and, most recently, wheat, have been genetically transformed for glyphosate tolerance. Although glyphosate has been used as an herbicide for 26 years, transgenic glyphosate-resistant crops are a more recent development and are widely deployed on acres devoted to soy and cotton. Research is underway to create other glyphosate tolerant crops. To date, two weed species, annual rigid ryegrass and goosegrass, have built resistance to glyphosate.

Corn, soy, rice, sugar beet, sweet corn and canola have also been genetically modified to tolerate the herbicide glufosinate ammonium. The seeds for these crops are sold commercially under brand names such as Liberty Link®. Transgenic soybeans, cotton and flax with a tolerance to the herbicide sulfonylurea are also on the market. Other strains of engineered soybeans and corn are resistant to sethoxydim, the active ingredient in the commercial herbicides Poast, Poast Plus, and Headline, used to control undesirable grass species. The herbicide bromoxynil, sold under the commercial name Buctril, is normally toxic to cotton, a broadleaf crop, and is primarily used on grass-like crops, such as corn, sorghum and small grains, to kill invading broadleaf weeds. Scientists have genetically modified cotton plants for resistance to this herbicide, allowing its use to control broadleaf weeds in cotton fields.

Improved nitrogen utilisation

There appear to be relatively few biotechnology applications specifically designed to enhance the characteristics of farm crops, such as size, yield, branching, seed size and number. Scientists have, however, created some enhancements. A recent example is the discovery of a gene in the alga Chlorella sorokiniana that has a unique enzyme not found in conventional crop plants. The enzyme, ammonium-inducible glutamate dehydrogenase, increases the efficiency of ammonium incorporation into proteins. In some plants, it increases the efficiency of nitrogen use. The practical implication is that less fertilizer would be necessary to grow these plants. When the gene was incorporated into wheat, biomass production, growth rate and kernel weight all increased, as did the number of spikes in the plant.

Stress tolerance

Stress tolerance involves a family of genes, rather than a single one. They are rapidly activated in response to cold, inducing the expression of "cold-regulated" genes, and resulting in enhanced freezing tolerance. Over-expression of these genes in Arabidopsis-small plants of the mustard family that are commonly used to study plant genetics-increases freezing tolerance and leads to elevated levels of proline and total soluble sugars, substances that protect against cold. Common stress responses in plants involve water retention at the cellular level. As a result, researchers have given special attention to osmoprotectant molecules, or molecules that hold water, such as sugars, sugar alcohols, certain amino acids (proline) and quartenary amines like glycinebetaine.

Various plants genetically engineered for increased levels of protectant sugar have shown increased drought tolerance. For instance, Arabidopsis and tobacco plants engineered to produce mannitol, a sugar alcohol, withstood high saline conditions and had enhanced germination rates and increased biomass. Other strategies have addressed different stress factors. Improved cold tolerance and normal germination under high salt was reported in Arabidopsis engineered to express the enzyme choline oxidase. Transgenic rice, engineered to express the late embryogenesis abundant protein gene transferred from barley, was significantly more tolerant to drought and salinity than conventional varieties of rice. Another transgenic rice engineered in the laboratory for enhanced expression of the enzyme glutamine synthetase had increased photorespiration capacity and increased tolerance to salt. Preliminary results suggested enhanced tolerance to chilling as well.

Regulation of Plant Hormones

Plant hormones such as auxin, cytokinins, gibberellins, abscisic acid, ethylene, etc have been targeted for genetic modification to influence plant growth and development; fruit development and ripening; stem elongation and leaf development; germination, dormancy and tolerance of adverse conditions. These hormone classes are highly interactive; the concentration of one affects the activity of another. For example, the ratio of the hormone abscisic acid to gibberellin in a plant determines whether a seed will remain dormant or germinate.

Recent discovery of an enzyme involved in the production of the hormone auxin enabled researchers to investigate the effects of moderating auxin production in determining plant characteristics. When auxin is overproduced, branching is inhibited and leaves curl down as the plant elongates, a reaction typically related to reduced light exposure. The same gene that produces this enzyme is apparently related to a gene in mammals that governs enzymes that detoxify certain chemicals.

In wheat, the hormone abscisic acid slows seed germination and improves the tolerance to cold and drought. Extending or enhancing the production of abscisic acid may also delay germination, a useful characteristic in climates where spring rain is sparse or falls late in the season. Production of abscisic acid is increased in response to environmental stress, and a family of enzymes called protein kinases stimulates its production. Selecting plant varieties high in abscisic acid, or engineering plants to produce more of the hormone, may confer greater drought and cold tolerance.

Introduction of dwarfed, high-yielding wheat contributed to the 'Green Revolution' of the 1960s and 1970s, during which world wheat yields almost doubled. Shorter varieties of wheat grains, with a greater resistance to damage by wind, resulted from a reduced response to the hormone gibberellin. Scientists have since shown that the gene called Rht can cause "dwarfing" in a range of plants, opening up the possibility of quickly developing higher-yielding varieties in several crops. Researchers believe that this strategy could be applied to a still wider range of crops through genetic engineering.

The plant hormone ethylene regulates ripening in fruits and vegetables. Controlling the amount and timing of ethylene production can initiate or delay ripening, which might reduce spoilage that can occur between the time produce is picked and brought to market. Transgenic techniques aim to regulate the enzyme that breaks down a precursor of ethylene production. By regulating the timing and rate of this degradation, ripening can be controlled. This technology has been applied and field-tested in tomatoes, raspberries, melons, strawberries, cauliflower and broccoli, but has not yet been commercialized.

Characteristics of Genetically Modified Crops (GMC)

Plant genetic engineering is aimed to increase nutrition, modify allergens and improve various functional attributes for farmers. Many genetically modified crops inclduign rice and wheet has received considerable publicity for its potential benefit to developing nations, where nutrient deficiencies are responsible for widespread health problems.

Increased Nutrition

Using bioengineering, scientists have added or modified nutrients in various crops, and created several nutritionally enhanced products. Although few have reached commercialisation, examples include adding iron to rice, or increasing beta-carotene and vitamin E in vegetable oils to boost the nutritional value. Other genetic modifications have altered the fatty acid composition in oils from soy and canola to create healthier fats.

Plants have also been engineered to increase phytonutrients-substances exclusive of nutrients that have benefits for improving health or preventing disease. These include isoflavones in soy and lycopene in tomatoes. Two genetic modification strategies have also been devised to increase the iron levels in cereal crops. One is the introduction of the gene that encodes for ferritin, an iron-storage protein. Over expression of this gene improves the storage capacity of plants by as much as three-fold. Using this and other genetic technologies, rice was engineered to contain beta-carotene, which it normally lacks, and enhanced iron content. This transgenic "golden rice" has yet to be bred into hybrid and native strains, so field testing of modified local varieties, commercial production and acceptance are still years away.

Another method for enhancing iron is reducing phytic acid content, which improves the degree and rate at which iron and other minerals are absorbed. In one experiment, corn genetically modified to be low in phytic acid was processed into tortillas. The iron absorption from these tortillas was 49 percent greater than from tortillas made with conventional corn. To further explore the effectiveness of iron absorption by reducing phytic acid, additional iron was added in the form of iron salt supplements and consumed with either strain of corn fed as porridge instead of tortillas. In this case, no absorption effect was observed. Although it is not clear why the phytic acid level had no effect, it is well known that when dietary iron levels increase, absorption decreases. Other substances in the diet may also have contributed to the reduced absorption.

While plants are the primary dietary source of vitamin E, they contain relatively low concentrations of the vitamin. Recent genetic engineering technology has been able to increase the vitamin E content of oils. As it happens, many seeds have abundant levels-up to 20-fold more of gamma-tocopherol, the immediate precursor of alpha-tocopherol, the active form of the vitamin. However, little of the gamma form is converted to the active vitamin. Researchers identified, isolated and cloned the gene responsible for expressing the enzyme that converts gamma-tocopherol to alpha-tocopherol. The gene was transferred to Arabidopsis, which subsequently exhibited a nine-fold increase in vitamin E.

Incorporation of this gene to stimulate similar gamma-tocopherol to alpha-tocopherol conversion into soy, canola and corn is probably not far in the future. Seed oils-particularly mustard and canola-have also been developed to contain carotenoids, especially beta-carotene, a nutrient widely studied for its role in cancer prevention. But this project is still in the testing stage. Protein (or rather specific amounts of essential amino acids, the building blocks of protein) is needed to fulfill human nutritional requirements for growth, health maintenance and muscle development. In regions of the world where cereal grains cannot be grown, people often rely upon starchy vegetables (roots, tubers or rhizomes) to supply most of their calories. While such crops often have high yields, the primary disadvantage is their very low protein content, less than one percent.

Researchers are seeking to improve protein content and quality in vegetable staples such as cassava and plantain through changes in amino acid profiles. For example, a non-allergenic seed albumin gene was introduced into the potato to increase its protein content. Transgenic tubers had 35 to 45 percent more protein and enhanced levels of essential amino acids. Moreover, transgenic plants produced more tubers and a yield increase of 3 to 3.5 percent. Scientists have also altered soybeans for higher protein in tofu. In an attempt to create healthier fats, researchers have modified the fatty acid composition of soy and canola in several ways. They have produced oils from soy and canola with reduced or zero levels of saturates; canola with medium chain fatty acids; high stearate canola oil free of trans-fatty acids; high oleic acid soybean oil, and canola with the long chain fatty acids gamma linolenic and stearidonic acid. The latter is of interest as an indirect source of docosa-hexaenoic acid (DHA), one of two long chain Omega-3 fatty acids shown to be beneficial in protecting against heart attack. DHA is available almost exclusively from seafood, primarily fatty fish. The plant precursor of DHA, linolenic acid, is poorly converted to DHA.

Transgenic high oleic acid soybean oil has 80 percent more oleic acid, one-third less saturated fatty acid than olive oil, and no trans-fatty acids. Researchers have also modified sunflower oil for high oleic acid content. Another type of modified soybean oil is low in saturated fatty acids and richer in linoleic acid than commodity soybean oil. Still another has reduced linolenic acid and no trans-fatty acids, increasing its stability for use as an ingredient in processed foods. Another seed unique for its high level of a single fatty acid is mangosteen. This tropical tree, grown in India, the East Indies and Southeast Asia produces seeds with as much as 56 percent by weight of stearic acid, a saturated fatty acid widespread in foods.

Stearic acid is noteworthy from a nutritional perspective for its stability and textural properties and because it is one of the few saturated fatty acids that does not appear to raise blood cholesterol levels. Thus, it is useful in fats for manufactured and processed foods. Enzymes cloned from mangosteen have also been expressed in canola with resulting increased levels of stearic acid. This research demonstrates the potential of the technology and the unusual sources of enzymes to alter fatty acid profiles in popular food oils such as canola.

Plant biotechnology has also aimed at increasing phytonutrients-substances in plants-exclusive of nutrients, that have benefits for improving health or preventing disease. For example, new research in nutrition suggests lutein may support multiple lines of defence against eye disease, and that lycopene serves as a powerful antioxidant in cancer prevention. Also called "accessory health factors" phytonutrients include isoflavones in soy, lycopene in tomatoes and polyphenols in green tea. In the laboratory, scientists have engineered tomatoes with 2.5 times as much lycopene as traditional tomatoes. At least one company is developing soy with more isoflavones, and canola with increased antioxidants and beta-carotenes, lutein and lycopene.

There are major constraints on this research, in part because there is still much about phytonutrients that is unknown. For example, some members of a class of phytonutrients may have deleterious effects while others are beneficial, as is the case with various flavonoids, water-soluble plant pigments that, while not considered essential, help maintain overall health as anti-inflammatory, antihistaminic and antiviral agents. In addition, scientists do not fully understand the biosynthetic pathways, or the succession of enzyme activities, for many phytonutrients. Another constraint is the limited scientific information about the safety and efficacy of potentially beneficial phytonutrients.

Some plants, especially cereals and legumes, are nutritious foods and feeds but also contain varying amounts of substances that interfere with digestibility and nutrient absorption. In excess, these materials may even be toxic. Genetic modifications are being explored to reduce these anti-nutritional substances, including phytate in cereals and legumes; glycoalkaloids such as solanine and chaconine in potatoes; tomatine, solanine, lectins and oxalate in tomatoes and eggplant; gossypol in cottonseed; trypsin and other protease inhibitors in soy, and tannins and raffinose in legumes. Phytate is widely distributed in cereals and legumes and reduces the absorption of iron, zinc, phosphorus and other minerals in humans and other animals. Phytate is indigestible for swine and poultry because their digestive tracts lack the enzyme phytase, which releases phosphorus from phytate.

Studies have shown that including phytase in the food ration improves phosphorus absorption and reduces phosphorus excretion. In the food animal industry, particularly for swine and poultry, high phytate feeds are associated with high levels of phosphorus excretion. Excess phosphorus in animal manures can be washed into streams or leach into ground water and become a serious source of water pollution. Research has indicated that poultry have substantially reduced phosphorus excretion when fed phytase as a supplement alongside ordinary soybeans or alternatively, genetically transformed soybeans expressing the phytase enzyme. Similarly, swine fed low-phytate corn showed increased phosphorus retention and reduced excretion. Genetically modified low-phytate corn contains at least five times as much available phosphorus as unmodified corn. Low-phytate corn feed was also associated with improved growth and finishing characteristics. In wheat engineered to expresses the enzyme phytase, seeds exhibited a two to four-fold increase in phytase activity. This opens the possibility of improving the digestibility of wheat, especially among non-ruminant animals.

Scientists are also seeking ways to reduce toxic substances such as glycoalkaloids. Researchers inserted antisense genes into potatoes to block the activity of the enzyme UDP-glucose glucosyl transferase, key to the production of the glycoalka-loid alpha-chaconine. This toxic substance can, at high enough levels, cause irritation of the gastrointestinal tract or impairment of the nervous system. Preliminary findings indicated that the transgenic potatoes produced fewer glycoalkaloids.

Allergens Modification

Food allergies and sensitivities cause a wide variety of conditions, symptoms and diseases, a few of which can be life threatening. A food allergy or hypersensitivity is one that provokes an immune response, while a food intolerance incites an abnormal physiological reaction. Experts estimate that 2 percent of adults, and from 2 to 8 percent of children, are truly allergic to certain foods. Food intolerance is a much more common problem than allergy. Unlike allergies, intolerances generally intensify with age. The eight most commonly allergenic foods are milk, eggs, peanuts, soybeans, fish, crustaceans, tree nuts and wheat. There are also significant allergies to non-food plants, such as ryegrass and other plants with airborne pollens that may cause hay fever or other seasonal allergic symptoms.

Most known allergens in food are proteins, suggesting the possibility of modifying the structure, or possibly eliminating the allergenic protein from the food. In some cases, traditional plant breeding has identified hypoallergenic strains that are targets for further genetic modification to reduce allergenicity. Neutralizing the allergens in major food grains would have an enormous impact on millions of families, where one or more members cannot eat these foods that are household staples. Researchers have used this approach in rice, the first food crop with reduced allergenicity to be created through genetic engineering. Further testing and development work continues to assure that people with known allergies to rice products can consume this genetically engineered food with-out developing their typical allergic reaction. In foods such as peanuts, however, which are highly allergenic to some sensitive individuals, the allergenic proteins constitute the majority of the plant's protein, so that elimination may not be possible.

Another example where genetic modification may be used to reduce allergenicity is in wheat, one of the "big eight" allergenic foods. Although not yet commercially available, scientists have genetically engineered wheat to overexpress the gene responsible for the synthesis of thioredoxin, an enzyme that catalyzes the reduction of disulfide bonds within protein molecules, thus reducing the protein's allergenic properties. When expressed in wheat, the enzyme reduced the bonds in the major allergenic proteins-the gliadens and glutenins-and to a lesser extent the minor ones, too, making them markedly less allergenic. At the same time, the functional characteristics of the wheat were not impaired.

Scientists are also exploring the potential of recombinant DNA technology to reduce the allergenicity of non-food allergens. For example, ryegrass is a dominant source of airborne pollen in temperate climates, and using antisense technology, scientists engineered ryegrass with reduced Lol p 5 protein levels. As this is the major allergen in ryegrass, the modification reduced the plant's allergenicity. Although genetic engineering has the potential to reduce allergenicity of foods, it also has the potential for unintentionally introducing new allergens.

Improved Functional Properties

Researchers are in search of enhanced functional properties for specific purposes, such as firmer tomatoes for canning, or beans with less breakage. One of the first applications to reach the market was the highly publicised Flavr Savr tomato, which was genetically engineered for delayed ripening. While the transformation process did delay ripening and extend shelf life, the product was expensive to produce and purchase and some consumers did not like the taste. This led to its withdrawal from the market. Other work aims to create a tomato that ripens on the vine but remains firm during harvest, handling and shipping. Firm tomatoes are preferable for canning, which consumes the largest share of tomato production. Using antisense technology, researchers have created tomatoes that are 40 percent firmer than their conventional counterparts and stay firm for at least two weeks.

Scientists have also engineered beans for desirable canning characteristics such as firm texture and seed coats that do not split. Several experiments are being conducted with soybeans. One would diminish that undesirable by-product of bean consumption, flatulence, by creating high sucrose soybeans through reduction of the carbohydrate raffinose. Another seeks to modify soybean oil to reduce the linoleic acid content so that it is more stable for industrial applications. Presently, barley is an unsuitable feed for poultry because poultry lack the enzyme to break down ß-glucan, the predominant polysaccharide (a type of carbohydrate) in endospem cell walls. Scientists have created transgenic malt that can depolymerise ß-glucan. Adding transgenic malt to barley-based poultry feed enabled poultry to metabolise barely, grow as well as poultry fed a corn-soybean diet, and produce more hygienic droppings. The digestibility of feeds can also be improved with modification of starch levels in different crops. For example, cattle can more readily digest amylose-free wheat in feed.

Extensive research has been directed toward altering the properties, quantity and distribution of starch in many plants, for a variety of purposes. The principal forms of starch are either linear (amylose) or branched polymers amylopectin. Using genetic engineering technology to influence the amount and length of chain branching and polymerisation increases the availability of starches with different properties. It also enables the development of novel starches. However, plants differ widely in where they store different types of starch; thus modification of starch production must be tailored to the particular plant. What works in the potato, for example, may not work in wheat or rice. Moreover, results in one variety of a crop may not be obtained in another. A well-known example of the modulation of starch synthesis has been the development of transgenic potatoes engineered to contain a gene for an enzyme affecting starch synthesis. The transgenic potatoes had up to 60 percent more starch than non-engineered strains. The increased starch content made the potatoes take up less fat during frying, resulting in a lower-fat product. About 40 percent of tapioca starch is used for the production of modified starch, sweeteners and the flavour enhancer monosodium glutamate. In processing tapioca, a significant amount of starch remains in the waste material and wastewater. It is estimated that even after extraction, the waste still contains 50 percent starch.

Bioengineered improvements in tapioca, such as reduction in water content and higher starch concentration, may increase the ease of processing the plant material into a finished starch product. Further, raising the efficiency of starch utilisation in the processing of sweeteners reduces the amount of starch reaching the waste stream. The possibility of converting the starch content of wastewater to energy, using high rate anaerobic digestion, is promising. However, a number of factors remain to be overcome, including the effect of environmental sulfates in the waste stream and the efficiency of energy production. The use of transgenic organisms offers potential solutions. When the enzyme thioredoxin is overexpressed in barley endosperm, the activity of the enzyme pullulanase, a rate-limiting enzyme in breaking down starch, increases four-fold. Breaking down starch is a key part of the barley malting process, and tests with this engineered variety showed that the time required could be reduced by up to a day. Over expression of thioredoxin also hastened barley germination, of special interest to growers of this normally slow-germinating grain.

4

Genetic Modification of Fruit Crops

Both plant cell and tissue culture, and more recently, plant molecular biology, have made available techniques that will greatly assist in the evaluation, collection and storage of plant genetic material. Much of the research into biotechnology has been carried out in industrialised countries, resulting in the comparative neglect of regional crops in developing countries, where many of the tropical and subtropical fruit species are grown. In addition, since the task of establishing collections of the world's plant genetic material began, most activity has occurred, quite naturally, with the major food crops. Very little effort has been applied to the minor crops and their wild relatives.

At the Commission on Plant Genetic Resources for Food and Agriculture in June 1995, the decision was taken to broaden the mandate to cover all components of biodiversity of relevance to food and agriculture. The Commission had already recognised in April 1991, that new biotechnologies should be applied to major local crops of social and economic importance, but of little international market importance and that protocols worked out for temperate species may not apply to tropical species.Tropical and subtropical fruit species should benefit from both of these decisions. Many of these fruit crops are vital to local communities in terms of variety of available food, nutrition, and therefore quality of life. In addition, there exists potential to bring some of these crops to an expanded international market by applying new biotechnologies, particularly those which can enhance flavour and postharvest characteristics. In recent years some of the minor tropical fruits have already become more popular, resulting in increased exports to countries such as USA, EC, Singapore and Hong Kong.

Tropical and subtropical fruit species are currently somewhat neglected as far as both germplasm collection and application of biotechnology are concerned. From a positive perspective, they represent a wealth of as yet untapped potential for valuable research effort and development. There is much that can be done using biotechnology techniques not only to conserve germplasm of these species but also to develop many of them as major fruit crops. There exists in the field of plant biotechnology the prospect of some spectacular breakthroughs, for example, being able to construct and store individual genes as genomic DNA or DNA sequence information. It is proposed by some that traditional methods of plant breeding and germplasm collection will be superseded. This possibility is currently a long way from reality and both new and traditional technologies have unique and complementary roles to play in the evaluation, conservation and use of plant genetic resources.

The survival of mankind has always depended upon the use of the earth's resources. It has been only in recent decades, as these resources have come under threat from a rapidly increasing population and from natural and manmade disasters, that the world generally has started to focus on the notion that we who take and use may be in some way responsible for the ongoing care and management of these precious resources. Since food and its production is one of the essential ingredients of mankind's existence, conservation of our resources is a particularly vital concept for those working in the field of plant science. There is evidence that as early as Egyptian times mankind has sought to select and conserve crop species with superior genetic qualities. Over the centuries there has been a natural ongoing process of crop improvement as production has been threatened or even devastated by pests, diseases or other cultural and socio-economic factors.

Our world environment is not static and so mankind has learnt to 'adjust' food production by exploiting the genetic diversity that occurs naturally. Genotypes which by virtue of their genetic composition are able to flourish in adverse conditions, whether in the cold, or in salty conditions or in the presence of insects or diseases, have been introduced or crossed with other lines to overcome problems in production. In the twentieth century this process has become a science as plant breeders equipped with greater knowledge have set about systematically developing elite lines and maximising crop yields. With the advent of plant biotechnology and the rapid development of gene transfer techniques, the potential to introduce desirable character traits is no longer restricted to those occurring in close relatives. In theory, genes, once identified, isolated and cloned may be introduced to a plant from any organism.

There are still many constraints and limitations involved in this process, which may be overcome in the future, and so the need for the conventional approach is not negated. Rather, the entire discussion serves to bring into sharp focus the value of genetic diversity as a natural resource and highlights the need to value and conserve it for future use. The escalation in the world population and the implications this has for food production has been a matter of concern for some time. It is probably the single most urgently motivating factor when considering the need for germplasm conservation. In 1992, the Convention of Biological Diversity made a foundational move to establish awareness in this area by focussing on the conservation of genetic diversity and the universal access to plant genetic resources, while acknowledging sovereignty over genetic resources. The principal reason for doing so was stated as being the need for increased productivity to feed current and future populations.

Germplasm Conservation in Fruit Species

Genetic resources are the raw materials which scientists use to alter plant performance, and with which they hope to achieve the dramatic increase in crop production that will be required in the decades ahead. Another factor of concern when considering biodiversity has been the emphasis on increasing yields by large scale production of superior clones. An obvious consequence of crop uniformity and neglect of wild related species, which often contain many useful traits, is a narrow gene pool and resultant vulnerability.

Papaya (*Carica papaya*) is an excellent example of a crop that is extremely vulnerable, but which is economically important in many tropical countries. In some of these countries, for example The Philippines, many farmers are already struggling with a subsistence existence and so the loss incurred by the failure of a cash crop causes real hardship. The crop is threatened worldwide by a number of serious diseases, including viruses, namely Papaya Ringspot Virus (PRSV-P), fungal diseases including *Phytophthora* root, stem and fruit rots, and mycoplasmas. PRSV-P has devastated crop production in many regions and may cause many landraces or useful genotypes to be lost completely. Extensive conventional crop breeding has achieved significantly improved fruit size and flavour and has been responsible for large improvements in crop yields but has been unsuccessful in addressing the current disease threats. However, the wild *Carica* relatives contain a huge pool of useful genes and valuable natural resistances to major papaya diseases have been observed. *Carica cauliflora, C. quercifolia, C. pubescens* and *C. stipulata* have resistance to PRSV-P; *Phytophthora*-resistance has been found in *C. goudotiana*; and *C. parviflora* appears resistant to two mycoplasma diseases in field plantings at Redlands Research Station in Southeast Queensland, Australia.

Although the focus in collecting germplasm has been placed on plant species that are valuable as food sources, the case of papaya illustrates the value of other genetic resources. Documentation of wild relatives, landraces and their useful characteristics and in-situ or ex-situ collection requires much time, effort and funds. However, the final benefits may be substantial, particularly as useful traits, such as resistance to a particular disease, potentially can be transferred to more than one crop species by gene transfer technology. The importance of germplasm conservation is being increasingly realised and many collections already exist.

The report on the State of World's Plant Genetic Resources records some of the difficulties being encountered, for example 'much of the plant genetic resources held in ex-situ collections are insufficiently and/or poorly documented' and 'globally, governments and donor agencies have made insufficient provisions for on-going maintenance costs of conservation infrastructure' resulting in a 'steady deterioration of many facilities and their ability to perform even basic conservation functions', and, 'a high percentage of accessions that are held in these banks are in need of regeneration'. These and other problems led to the adoption of The Global Plan of Action at the fourth International Technical Conference on Plant Genetic Resources for Food and Agriculture in June 1996.

Any attempt to systematically conserve genetic material on a worldwide basis will of course necessitate extensive government cooperation and funding. The task at hand is a daunting one but is complicated by the prevailing attitude of many governments in western countries to agricultural research. Although they are signatories to all relevant conventions there is a growing reluctance to fund agricultural research. As funds dry up, competition causes available funding to go to those research areas promising significant results in the short term. Germplasm storage is not offering spectacular short term outcomes and is motivated by the need to take responsibility for the long term. All countries need to value and be responsible for the conservation of genetic diversity. It is a natural resource, the value of which is still to be fully appreciated as science continues to unravel the mysteries of plant function at a cellular level and employ them to meet the constant stream of new and ongoing pressures and difficulties facing crop production. There is an urgent need to increase research into technologies that facilitate conservation of plant genetic resources. This is consistent with The Global Plan of Action.

Since the 1970's there has been a conscientious effort to collect germplasm. In 1995, Tao and Anishetty reported that worldwide more than 4 million accessions of crop plants and their relatives were stored in genebanks and that there were 256 genebanks with long-term storage facilities and 1227 institutions with ex-situ collections. However, these statistics are rendered less impressive by two considerations. Firstly, 40% of these accessions are cereals (wheat represents 14%). No consistent effort has been applied to the collection of other major crops, while minor crops including fruit crops, have been largely neglected. Secondly, more than 90% of the accessions are stored as seed. As a consequence, vegetatively propagated species and species with recalcitrant seeds are greatly underrepresented. In addition, if we are to preserve diversity, collections must also include wild relatives of crop species. As gene transfer techniques overcome the incompatibility between species, there is also a need to be mindful that non-crop species may also represent a valuable source of useful genes such as disease and insect resistance.

Although seed storage has been the conventional method of conserving germplasm, pollen storage and field collections have also been employed. Scientific advances in the field of cell and tissue culture systems have made available, as alternatives, conservation *in vitro* in either slow growth systems or cryopreservation. More recently, developments in molecular biology offer the future prospect of DNA storage as libraries of either DNA sequence information or of DNA constructs. As reported in the State of the World's Plant Genetic Resources, one of the major concerns with germplasm storage is the maintenance of collections. Stored germplasm requires careful characterisation as well as periodic regeneration to ensure viability. Techniques resulting from biotechnology offer improvements in this area. Molecular marker techniques enable characterisation at the DNA level, and an advantage of *in vitro* conservation is that viability is easily assessed visually. As regular regeneration, usually annually, is essential, it is less likely to be overlooked. A major effort is still required in the evaluation and conservation of tropical and subtropical fruit species.

The development and maintenance of germplasm banks is of limited usefulness if the genetic material is not distributed and used worldwide to develop new and improved varieties of food plants. Hence The Global Plan of Action promoted the development of networks for the effective exchange and utilisation of germplasm. It states that 'networks are important platforms for scientific exchange, information sharing, technology transfer, research collaboration, and for the determination and sharing of responsibilities for such activities as collecting, conservation, distribution, evaluation and genetic enhancement'. FAO has supported and collaborated in the development of networks in the 1990's. Networks have been established on both a crop (mushroom, cactus pear, olive, citrus, nut tree) and regional basis. The innovative aspect of the newly launched networks is the promotion of a coordinated approach to identifying, evaluating and conserving the genetic variability of selected crop species, with the aim of its utilisation for the improvement of cultivars and their adaptation to farmer's needs'. There exists the potential to strengthen existing networks, develop new networks and to coordinate these regional and crop-based networks into an effective global system.

The chief aims of The Global Plan of Action are the conservation and sustainable use of genetic resources, by developing more effective conservation strategies at both the national and international level. A major advantage of crop-based networks is that they can facilitate linkages between conservation and sustainable utilisation of crop genetic resources. A number of factors highlight the importance of a global approach. Plant Genetic Resources (PGR) are a common concern of humanity and all nations are both donors and users of PGR. The effective use of a wide range of PGR in all countries is an incentive to ensure their conservation and maintenance. Thus the international community as a whole has both an interest in and a responsibility to conserve PGR and this is the basis for an effective integrated and rational Global Plan of Action.

A related development which emphasises the need for a global approach is the proposed World Information and Early Warning System on Plant Genetic Resources (WIEWS). The technical meeting on the methodology of WIEWS recommended the establishment of a global network, on a voluntary basis, under the auspices of FAO. WIEWS would provide a vehicle to register cases of genetic erosion both in *in situ* populations and *ex situ* collections. However, it would require an accurate information base on the current status of PGR including crop species, local varieties and wild relatives. The Global Network would provide 'timely and precise information on PGR conservation and utilisation', contribute to the implementation and updating of The Global Plan of Action, provide an information flow between countries and develop an effective early warning system.

In vitro Regeneration of Plants

Micropropagation has been defined as '*in vitro* regeneration of plants from organs, tissues, cells or protoplasts' and 'the true-to-type propagation of a selected genotype using *in vitro* culture techniques'. True-to-type propagation has important benefits for plants that have no named varieties, such as Australian dioecious papaw genotypes where traditional plant breeding has failed to produce stable lines. Micropropagation has also been useful for the rapid initial release of new varieties prior to multiplication by conventional methods, eg. pineapple and strawberry, and germplasm storage for maintenance of disease-free stock both in controlled environment conditions and long term via cryopreservation. However, the ability to propagate plants *in vitro* free of genetic off-types is dependant on the technique used for micropropagation. Failure to understand this principle has resulted in disastrous consequences with some species, for example the commercial production of banana clones containing 90% dwarfs.

Protocols which have been developed for *in vitro* propagation of plant species can be divided into three systems. System 1 is based on callus culture followed by organogenesis or embryogenesis. System 2 comprises proliferation of axillary buds and/or adventitious buds, resulting from repeated subculture on multiplication media containing cytokinin. In system 3, micro-cuttings obtained from axillary buds of apically dominant shoots are grown on hormone-free or low cytokinin media. System 3 has been used in the author's laboratory for clonal propagation of papaw, neem, coffee and passionfruit and is not prone to production of genetic off-types.

Exploitation of the variation that exists in populations has led to the development of many commercial varieties and hybrids. With the rapid expansion of tissue culture technologies came the observation that genetic variation was occurring in plants regenerated from somatic cells and this was seen as a novel source of variation. The frequency of off-types varies with species, culture type and number of sub-cultures and has been attributed to a number of variations within cultured cells. However, the cause of somaclonal variation is not fully understood; therefore, it cannot be controlled and changes can be epigenetic and unstable. Variation can be promoted by the use of radiation or chemical mutagens and the use of colchicine to change ploidy. The potential of somaclonal variation involves the ability to change one or a few characters without altering the remaining part of a genotype.

Initial notes of potentially useful variation included increases in cane and sugar yield, and resistance to eye-spot disease in sugar cane; improved tuber shape, colour, and late blight resistance in potato; and, increased solids and *Fusarium* resistance in tomato. *In vitro* selection involves the screening of cell cultures that are exhibiting genetic variation, for tolerance or resistance to pathogens, herbicides, low or high temperatures, metals and salt. It necessitates a reliable method of regeneration from callus. For successful application of the technique, tolerance at the cellular level must also occur in the regenerated plant in the field and be transferable to other plants via conventional plant breeding.

The use of *in vitro* screening to select for disease resistance is most effective for diseases which produce toxins. This was initially demonstrated by Carlson who regenerated tobacco plants resistant to *Pseudomonas tabaci*, from haploid cells resistant to methionine sulfoxamine, an analogue of the disease toxin. Similar successes have been noted with a range of diseases including *Phoma lingam* in canola, *Helminthosporium oryzae* in rice, *Pseudomonas* and *Alternaria* in tobacco and *Pseudomonas solanacearum* in tomato. Addition of herbicides to culture is an ideal system for *in vitro* screening as defined concentrations can be used.

However, effectiveness is dependent on the mode of action of a herbicide on the whole plant being similar at the cellular level, and is therefore better suited to herbicides that interfere with basic metabolic functions. Resistance or tolerance has been developed *in vitro* to a range of herbicides including paraquat in tomato, picloram amitroie and triazine in tobacco and 2,4-D in *Lotus corniculatus*. Carrot suspension cultures which were resistant to 35mM glyphosate in solution were selected after 8 subcultures on low concentrations of glyphosate (0.3 to 0.6 mM). Carrot plants regenerated from these cultures were resistant to glyphosate sprays in field plantings. A similar stepwise selection procedure for gene amplification produced tobacco cell lines with high tolerance to glufosinate.

Cell cultures that are resistant to increased salt (NaCl) *in vitro* do not necessarily produce regenerants which tolerate high salinity levels in soil. For example, the use of somatic cell cultures of rice did not lead to heritable salinity tolerance in the field; however, this was eventually achieved after *in vitro* selection of callus cultures initiated from immature ovaries and anther culture-derived lines. Examples of genotypes which had been released or were in field trials in the USA and Canada. Many of his references were patent applications or personal communications. These included chlorsulfuron and imidazilinone resistance in canola, imidazilinone resistance in corn, high solids and *Fusarium* race 2 resistance in tomato, a white flowered form of lucerne, potato virus Y resistance in tobacco and fall armyworm resistance in sorghum. Potentially useful somaclonal variants have been identified in a number of fruit crops. These include *Phytophthora* resistance in papaya and apple, *Xanthomonas* resistance in peach, *Erwinia* resistance in pear and salt tolerance and 2, 4-D tolerance in orange.

Embryo Culture Technique

Embryo culture was probably the first tissue culture technique to be applied to plant improvement. Applications of embryo culture are rescuing embryos after interspecific hybridisation, clonal propagation of families such as *Gramineae* and conifers which contain recalcitrant species, and overcoming seed dormancy and seed sterility. Interspecific crosses are often attempted to transfer desired traits such as disease resistance, stress tolerance or high yield from wild species into important crop species, eg. cotton, soybean and papaya. Incompatibility after these crosses normally results in embryo abortion and this is often caused by breakdown of the endosperm or embryo-endosperm incompatibility. *In vitro* culture of hybrid embryos often successfully bypasses postzygotic incompatibilities.

There are numerous references of embryo rescue of incompatible hybrids. Another approach to wide hybridisation is *in vitro* fertilisation of cultured ovules and ovaries. This was achieved initially with opium poppy. The major application of this technology in monocotyledonous crop species is in maize. In an extensive research program funded by the Australian Centre for International Agricultural Research (ACIAR) techniques have been developed for interspecific hybridisation and embryo rescue of papaya and wild relatives. Hybrids have been produced between *C. papaya* and *C. cauliflora, C. quercifolia, C. pubescens, C. parviflora* and *C. goudotiana*. Useful characteristics of the wild species include papaya ringspot virus resistance, *Phytophthora* resistance, high sugar content, and ornamental characteristics.

Gene Transfer Techniques

Many years of research have culminated in an increasingly detailed understanding of biology and genetics at the molecular level. Combined with tissue culture technology, this knowledge is being applied to the modification of plant genomes. Because of the universality of the genetic code, potentially useful genes can be transferred to plants from any organism. This has lead to the development of new and improved genotypes by the addition of single genes that code for traits such as insect or disease resistance, or by the inactivation of single gene faults. Future potential includes the possibility of transferring larger DNA constructs that code for multiple genes conferring more complex traits. Applications of gene transfer are increasing rapidly, driven by an extensive research effort worldwide. Initial successes were in the development of disease, insect and herbicide resistant crops.

More recently, gene transfer is resulting in control of plant development in individual species. Inhibition of expression of polygalacturonase in a tomato genotype lead to the first commercial release of a genetically engineered plant - the FLAVR-SAVR tomato. Extensive research is being directed towards gene inactivation techniques for control of ethylene production, and thus ripening, in climacteric fruits. Gene transfer techniques have also lead to the development of carnations with increased vase life and new colour forms. Isolation of a flower-meristem-identity gene from *Arabidopsis* has the potential to dramatically reduce time to flowering and hence length of the juvenile phase in plant species.

Developing applications include transformed plants which will produce more or increased levels of valuable oils, proteins and starches, and plants that produce vaccines and can be used for oral immunisation against a range of serious human diseases. Useful agricultural applications of gene transfer necessitates incorporation of a foreign DNA construct into a plant genome, the regeneration of transformed plants, the stable expression of the introduced gene and inheritance in subsequent generations. The successful application of plant transformation techniques is dependent on plant tissue culture protocols to regenerate transformed plants. Since the first reports of gene transfer with species that are relatively easy to tissue culture, petunia and tobacco, transformation procedures have been published for a wide range of species including tree crops, such as apples and papaws.

Transgenic plants of a number of fruit species have been produced, including kiwi-fruit, citrus, strawberry, grape, cranberry, peach and plum. Numerous DNA delivery systems have been noted. Direct gene transfer systems such as microinjection, electroporation and polyethylene glycol have been used to transform protoplasts. Protoplasts are easily damaged and regeneration of plants from them is very difficult with many species. Consequently, such systems are not widely used. The popular method of direct gene transfer is particle bombardment. DNA coated microparticles are bombarded into tissues, and major crop species have been transformed using this technique, for example monocots such as wheat, barley, rice and recalcitrant dicot species such as cotton, and soybean. The advantages of the particle gun in terms of tissue culture regeneration is that it can be used on callus and suspension cultures and on organised tissue of both monocots and dicots.

Increasing the number of options for tissue culture technology increases the probability of successful regeneration. The preferred method of gene transfer for dicots is *Agrobacterium* mediated transfer. Co-cultivation of callus, suspension cultures or leaf discs with *Agrobacterium* has been used to successfully transform many species, for example efficient systems have been described for major crop species such as potato, tomato and sugar beet. Virus, insect and herbicide resistance in a range of species including tobacco, potato, tomato, lucerne and soybean have been successfully trialled in field conditions and are in the early stages of commercialisation. Transformation of cereals, which are the world's most important food, has been by direct DNA uptake systems, particularly biolistics. Embryogenic cultures in cereals have generally been developed from immature embryos and inflorescences and subsequently transformed in the form of protoplast, suspension or callus cultures. The most recalcitrant of these species is wheat, where difficulty in developing reliable regeneration systems is exacerbated by low frequency of transformation.

Other problems associated with tissue culture of cereals are loss of regenerative capacity with increasing number of subcultures and occurrence of abnormalities such as sterility in regenerated plants. *Agrobacterium* was thought incapable of infecting monocots, however recent notes describe *Agrobacterium* mediated transformation of rice, maize and barley which offer considerable potential. Systems have been noted for transformation of haploids via particle bombardment of pollen and via microinjection of DNA into microspore derived embryoids of canola. More recently, maize has been transformed by pollen which was mixed with *Agrobacterium* containing binary plasmids. Such systems may provide a method that eliminates the need for *in vitro* regeneration, which is a major limiting step for some species and is prone to somaclonal variation.

Although in theory tissue of any species can be transformed, it is currently not possible to regenerate transformed plants at high frequencies for many species. Current methods of gene transfer and selection can be disruptive to growth of *in vitro* cultures, thus it can be difficult to develop efficient transformation procedures for species that are recalcitrant *in vitro*. In addition, there is a need to be able to deliver gene constructs to tissue that is competent both for transformation and regeneration. Limitations of current technology include low frequency of transformation, high frequency of undesired genetic aberrants, unpredicitability of transgene expression, the need for techniques to transfer large DNA sequences coding for multiple genes and the absence of efficient repeatable regeneration protocols for many species *in vitro*.

Molecular Markers and Gene Mapping

One of the recent applications of new techniques of molecular biology is the rapid development in molecular markers and gene mapping. The use of DNA-based markers is allowing researchers to mark genes or chromosome regions that are related to genetic traits such as disease resistance or fruit colour. A particular advantage of such techniques is that complex polygenic traits can be analysed. Mapping techniques can also be used to isolate genes based entirely on their genetic behaviour. Once these have been identified, sequenced and cloned, gene transfer techniques can be used to transfer them to other species.

Various techniques have been used to mark individual characters in segregating populations and construct genetic maps. Initially isozymes were used as markers. Isozymes are enzyme products of particular genes. Individuals which vary in their genetic characteristics are separated by polymorphisms in isozyme banding patterns on gel plates. Recently, the more precise marker techniques such as RFLPs, RAPDs, AFLPs and VNTRs are being preferred. Restriction Fragment length Polymorphisms (RFLPs) are short fragments of DNA generated by restriction enzymes. The fragments are separated by electrophoresis and are identified with radioactive probes. Random Amplified Polymorphic DNA (RAPDs) are produced by randomly amplifying DNA sequences using PCR technology and primers. AFLPs are a related PCR-based technique. Highly polymorphic markers which also depend on PCR technology are VNTRs. These are tandem repeats of short DNA sequences and include minisatellites and microsatellites. The use of RAPDs, AFLPs and VNTRs is much quicker than RFLPs. Using these techniques it is possible to link markers on a genetic map with traits of interest in a particular species.

Once markers are defined, they can be used directly in breeding programs since they allow researchers to predict phenotypes based on the presence or absence of the marker. Alternatively, once genes for particular character traits have been identified by marker techniques, they can be cloned. Although currently a difficult task, the development of automated systems will likely make such 'map-based cloning' fairly routine in the future. Major research efforts are underway to sequence, or partially sequence, all genes in model species such *Arabidopsis* or rice. Once this is achieved in a major dicot and monocot species, it will be easier to locate particular genes in other species. Genes that are located together in one species are likely to be similarly located in another species. Alternatively, a gene which codes for a similar function in a related species may be found by identification of homologous DNA sequences. Although much of the initial research effort has been applied to important vegetable and cereal species, these techniques will have a large impact on fruit tree breeding. For example, the long juvenile period required by many fruit tree species has made the process of breeding and selection of superior genotypes a long term effort. If a genetic marker is obtained for a trait of interest, plants could be selected at the seedling stage.

Genetic Modification of Tropical and Subtropical Fruits

Micropropagation

Micrografting, an innovative technique which should be applicable to other tropical and subtropical species, was developed in 1972 for *Citrus*. Micrografting of *Citrus* results in virus-free plants that are not juvenile. Most of this research is at a preliminary stage and this is demonstrated by two factors. Firstly, with a few exceptions (citrus, banana, papaya, pineapple), most of the successes have been achieved using juvenile bud explants, and in many cases seedling tissue. Secondly, only 50% of the reports document acclimatisation of plantlets in soil. However, the increase in research effort in this area in 1990's and published results on a wide range of tropical fruit species is encouraging and justifies a larger research effort.

Regeneration of plantlets from callus can follow two distinct pathways. Organogenesis involves the development of meristematic tissue which produces either shoots or roots, with vascular connections to the parental callus. Somatic embryos progress through characteristic developmental stages (globular, heart, torpedo and mature), have both a shoot and root meristem (bipolar) and become independent from the parental callus. As with most species, tropical and subtropical fruit species tend to regenerate by either organogenesis or embryogenesis, although occasionally there are reports of regeneration via both pathways (eg. *Carica papaya*). As with micropropagation, most of the published successes via regeneration from callus has been achieved using juvenile tissue explants.

Most tropical fruit species are woody perennial dicots, which in general are a difficult group to culture *in vitro*. Tissue culture of recalcitrant species has been achieved in many cases using immature embryos or very juvenile seedling explants, as these often represent the most regenerative tissue in a plant. Much of the documented research on recalcitrant species entails a thorough evaluation of a range of media components and concentrations. However, recent innovative protocols for embryogenesis and regeneration of *Musa* spp. from male flower explants demonstrate the importance of testing a range of explants as well as media when working with untried or 'difficult' species. However, few species are truly recalcitrant, and most can be regenerated provided they can be subjected to a concentrated research effort. *In vitro* culture of papaya has been researched extensively worldwide and highly refined protocols have been developed. Efficient methods for embryogenesis from immature zygotic embryos have been used to develop transgenic plants and encapsulated artificial seeds. Recent notes of culture of integuments of immature seeds and the development of embryos of maternal origin should have important implications for rapid propagation and production of transformed 'elite' papaya genotypes.

Regeneration of plantlets via embryogenesis following culture of nucellar tissue of mango, another recalcitrant species, demonstrates what can be achieved by innovative research. There are two other noteworthy achievements with regeneration of tropical fruit species. Plantlets have been produced after embryogenesis from mature leaf tissue of mangosteen, and seedless triploid pummelo plants were produced following culture of endosperm. Regeneration via organogenesis or embryogenesis is seen as having three major applications: micropropagation, cryopreservation and gene transfer. However, regeneration from callus can be prone to production of genetic off-types, thus any micropropagation technique should not be assumed to be clonal until large scale field testing has been undertaken to evaluate a protocol for genetic stability. Similarly, embryogenic cultures are well suited to cryopreservation studies, however an assessment of genetic stability after large scale field evaluation is also necessary for each species and protocol, before it can be endorsed as a method of germplasm conservation. Regeneration protocols are basic to the development of transformation systems, as transformed plantlets have to be grown from single transformed cells. Thus a large research effort is warranted for many tropical and subtropical fruits, if

Embryo Culture Techniques

Production of plants from mature embryos is one of the most straight forward *in vitro* techniques. Immature embryos represent the most regenerable tissue for many species. Consequently, culture of immature embryos has facilitated both *in vitro* regeneration of recalcitrant species and rescue of interspecific hybrids. The potential of embryo culture for the regeneration of recalcitrant fruit species has been demonstrated with *Litchi chinensis*. Lychee is one of the most difficult tropical fruit species to establish and grow *in vitro*. Kantharajah *et al.* cultured immature embryos of lychee as small as 3mm in length. Ability to produce adventitious buds varied with genotype, however cultivar 'Bengal' produced 15 shoots after pretreatment of immature embryos with BAP. Root formation was achieved with 65% of adventitious shoots, and resulting plantlets acclimatised.

Although protoplasts can be isolated from a range of tissues of almost any plant species, regeneration of plants from protoplasts is one of the most difficult *in vitro* techniques. Reports of success with recalcitrant woody species are limited, and applications to tropical and subtropical fruit species are rare. The only exception is with *Citrus* species. There have been notes of protoplast isolation of non-woody tropical fruit species: papaw and banana, although sustained cell division was not achieved.

The principal uses of protoplasts are as targets for direct methods of gene transfer and to facilitate interspecific hybridisation. Direct methods of gene transfer requiring protoplasts are becoming obsolete. Although protoplast fusion may provide a method of interspecific hybridisation, the difficulties in achieving sustained cell division and plantlet regeneration in many species generally outweigh the advantages. In reality, protoplast fusion is only an advantage when prezygotic barriers prevent interspecific hybridisation. If zygotic embryos are produced, embryo rescue and culture is a superior technique.

A comparison of these techniques has been demonstrated with interspecific hybridisation of *Carica papaya* and related species. An efficient method has been developed for embryo rescue, embryogenesis and plantlet production for hybridisation between *C. papaya* and *C. cauliflora* and between *C. papaya* and *C. pubescens, C. quercifolia, C. parviflora* and *C. goudotiana*. By contrast, attempts to hybridise *C. papaya* and *C. pubescens* have resulted in callus growth but no plants. Similar difficulties have been experienced in the authors laboratory when protoplasts of *C. papaya* were fused with those of *C. pubescens, C. quercifolia* and *C. stipulata*. As with most difficult aspects of *in vitro* culture, intensive funding and research effort usually result in success. Protoplast culture of tropical and subtropical fruit species is no exception and, given sufficient attention, new and improved protocols would probably be developed. However, at this stage, it would be preferable to direct funds towards other aspects of *in vitro* culture that appear to hold greater potential.

Application of Genetic Markers

The development of molecular markers is a rapidly expanding field. To date a major effort has been applied to model species such as *Arabidopsis,* and to major crops such as tomato, rice and maize. There has been some application of isozyme markers to tropical and subtropical fruit crops eg. *Musa* spp., mango, papaya and pineapple. By comparison there are few notes of the use of DNA markers in tropical and subtropical fruit species. Cultivar identification has been achieved by use of satellite probes in avocado, mango and papaya and by simple sequence repeats and minisatellite probes in avocado, lychee, mango and papaya. Magdalita *et al.* used RAPD analysis to confirm interspecific hybrids between *Carica* species. There is great potential for the application of genetic markers to tropical, subtropical and indeed all perennial fruit crops.

Firstly, DNA fingerprinting allows accurate identification and characterisation of genotypes and species.

Secondly, DNA markers provide a rapid method of analysis of genetic variation in a population. Analysis of genetic variation between *Carica papaya* and related species facilitated a breeding program to introgress disease resistance genes from wild relatives into *papaya*. Off-types, resulting from genetic variation in *in vitro* regeneration systems, can be identified by DNA markers in some instances. A RAPD marker has been developed to identify dwarf off-types in banana. Dwarf off-types have been a major problem in micropropagation of elite banana cultivars.

Thirdly, the use of molecular markers can facilitate plant breeding programs. Consequently, conventional breeding and assessment based on morphological markers can be a difficult and slow process with perennial fruit species.

A combination of *in vitro* techniques and use of molecular markers may help overcome many individual problems. Marker assisted breeding could be particularly useful for gene introgression, breeding for multigene resistance, and resistance to a disease not yet in a region or country. An example of the savings that could be made in terms of time and funds, is the potential to select, at the seedling stage, for recessive characters such as red female dioecious papayas. The advantages for fruit species with much longer juvenile phases is obvious. Currently, the high cost of this research is preventing large scale application in tropical and subtropcial fruit species. However, it represents a wide open field of research and the potential gains warrant a large investment of funds.

As with marker technology there is enormous potential in applying recombinant DNA technology to tropical fruit species. Gene transfer offers the potential to add one or two genes to an elite genotype. Most tropical fruit species are highly heterozygous. They often have difficulties that interfere with traditional plant breeding programs and have long juvenile periods. A good example is papaya industries that are based on dioecious female genotypes. In Australia, breeding programs over 20 years have produced elite female genotypes after years of concentrated effort, however the genotypes have not been maintained in future seed generations. Plant biotechnology offers the potential to add one characteristic (eg. Papaya Ringspot Virus Resistance, PRSV-P) to an elite genotype, which could then be micropropagated *in vitro*.

Papaya has become the model crop for application of transformation technology to tropical fruit crops. Transgenic papaws resistant to PRSV-P have been produced and are currently being tested in large scale field plantings in Hawaii. There are also a few preliminary notes on transformation of *Musa* spp., passionfruit and pineapple with marker and reporter genes. The only published notes on transformation of a woody perennial tropical fruit species are in mango.

Transformation of other tropical and subtropical fruit species is warranted and could result in considerable benefits. Potential includes reduced juvenility in tropical fruit species (eg. mangosteen), disease resistance (eg. *Phytophthora* resistance in durian) and insect resistance (eg. resistance to fruit fly). Recent advances in understanding and cloning of promoters provides the potential of tissue specific expression of genes, hence insect resistance could be targeted to meristems, roots, or epidermal tissue. There have been numerous recent notes on transformation of temperate woody perennial fruit species. This demonstrates that transformation of woody species is achievable. In addition, in the last decade, *in vitro* regeneration systems, which underpin transformation technology, have been developed for some tropical and subtropical fruit species.

Use of Plant Genetic Resources

Biotechnology can make major contributions, not only to crop improvement, but also to the conservation and use of plant genetic resources. The emphasis at sessions of the Commission on Plant Genetic Resources has been on the potential of biotechnology to contribute to increasing food production, to promote sustainable agriculture, to promote the conservation of plant genetic resources and biodiversity, and to effect free exchange of scientific information and plant genetic resources. Consequently, the Commission agreed to the need for a Code of Conduct to address

(i) the promotion of sustainable use of biotechnology in the conservation and use of plant genetic resources;

(ii) the promotion of access to plant genetic resources;

(iii) the promotion of biosafety to minimise environmental risks throughout the world; and

(iv) the equitable sharing of the benefits of biotechnology between the owners of the technology and the donors of the germplasm.

A draft international code of conduct for plant biotechnology was submitted to the fifth session of the Commission in April 1993. This session also recommended actions to maximise positive effects and minimise possible negative effects of biotechnology, including action by governments to establish national research programs in appropriate biotechnologies. Appropriate technologies were defined as being 'technically feasible, bringing tangible benefits to the users' while being 'environmentally safe, socio-economically and culturally acceptable'. The impact of biotechnology on crop improvement has been extensive, and its potential influence on the process of conservation of plant genetic resources and biodiversity is profound. The relationship between plant biotechnology and the conservation of genetic resources is a reciprocal one.

Biotechnology offers many novel techniques that will help maximise efforts to conserve genetic resources, while the conservation of genetic resources is vital for future research and development of biotechnologies. The world's genetic resources are the raw materials, not only for plant breeders, but also for developing biotechnology industries. This chapter provides an overview of biotechnologies that can contribute to the conservation process, followed by their specific application to tropical and subtropical fruit species and areas of future potential.

In vitro Techniques

In vitro collecting

In vitro collecting involves initial disinfestation and placement of plant explants in sterile culture medium, before transport to a tissue culture laboratory for further *in vitro* procedures. *In vitro* collection is particularly useful for species that are vegetatively propagated and for those with recalcitrant seeds or embryos which deteriorate rapidly. The technique has much potential to facilitate the collection of germplasm of tropical and subtropical fruit species, as has already been demonstrated with Cassava and coconut. Recently, 300 *Musa* accessions were collected in Papua New Guinea using this technique, before being transported to a collection in Australia. An added advantage of this exercise is that it complied with quarantine regulations that are in place to stop the spread of *Fusarium* and other diseases.

In vitro culture

In vitro culture offers some major advantages for conservation and use of plant genetic resources. *In vitro* cultures can be established from disease-free parent material and maintained in a disease-free state. Alternatively, meristem culture, in combination with treatments such as chemo- or thermotherapy, is a proven technique for elimination of specific viral and other diseases. Consequently, disease-free germplasm can be safely and rapidly transferred between countries. However, it is not safe to assume *in vitro* cultures are free of specific viruses, unless the initial parent material has been cleared by appropriate ELISA or equally stringent testing. Caution should always be exercised when dealing with export/ import of plant tissue cultures.

In vitro culture has been used to facilitate international collaboration in a research project between Australia and the Philippines on development of PRSV-P resistance in papaya. Sterile cultures of *Carica* species, interspecific hybrids and embryogenic cultures were regularly transferred between the two countries and contained in quarantine approved laboratories. For example, cultures from the Philippines have been transported to Australia, subjected to *in vitro* procedures in the authors laboratory and returned to the Philippines. Australian quarantine restrictions that prevent early field release of this material are not contravened.

In vitro germplasm storage

Ninety percent of all accessions held in germplasm collections are stored as seeds. There are three limitations of seed storage. Firstly, it is not applicable to species with no seeds eg. *Musa* species and *Artocarpus altilis* (breadfruit). Secondly, for highly heterozygous species, eg. dioecious papaya, it is preferable to conserve vegetative or clonal material. Thirdly, seeds of recalcitrant species are sensitive to desiccation and/or chilling, and tropical recalcitrant species often lack a natural dormancy mechanism, eg. *Mangifera indica* (mango), *Nephelium lappaceum* (rambutan) and *Cocus nucifera* (coconut).

In vitro storage also represents a viable alternative to field collections for species included in the three groups for rare and endangered species. Field collections require regular maintenance and are prone to damage from disease and insect attacks, extremes of weather and natural disasters. By comparison *in vitro* collections are safe from these problems, although failures in environmental control can result in loss of cultures. Thus duplicate collections are advisable. Another advantage of *in vitro* storage over field collections is that large numbers of cultures can be stored in a small volume in a growth room.

In vitro collections are generally maintained in either slow growth storage or by cryopreservation. Slow growth storage employs various methods to reduce sub-culture frequency, and thus labour and media costs. Cryopreservation involves long term growth suspension in liquid nitrogen. Cryopreservation was achieved initially using cryoprotectants and controlled cooling. This classic technique has been successfully applied to cell suspension and callus cultures. Newer methods involve dehydration of cultures before rapid freezing by immersion in liquid nitrogen. This method shows potential with organised tissue, including embryos and meristems. There exist a number of different dehydration/freezing procedures, including vitrification, encapsulation dehydration, desiccation, pregrowth and droplet freezing.

5

Development of Microbiology

A microorganism or microbe is an organism that is microscopic (invisible to the naked eye). Microorganisms are often described as single-celled, or unicellular organisms; however, some unicellular protists are visible to the naked eye, and some multicellular species are microscopic.

Microorganisms can be found almost anywhere in the taxonomic organisation of life on the planet. Unicellular organisms carry out all the functions of life. Bacteria and archaea are almost always microscopic, whilst a number of eukaryotes are also microscopic, including most protists and a number of fungi. Unicellular species are those whose members consist of a single cell throughout their life cycle. This qualification is significant since most multicellular organisms consist of a single cell at the beginning of their life cycles. Unicellular organisms usually contain only a single copy of their genome when not undergoing cell division, although some organisms have multiple cell nuclei.

Microorganisms are found in virtually every habitat present in nature. Even in hostile environments such as the poles, deserts, geysers, rocks, and the deep sea, some types of microorganisms have adapted to the extreme conditions and sustained colonies; these organisms are known as extremophiles. Some extremophiles have been known to survive for a prolonged time in a vacuum, and some are unusually resistant to radiation. Many types of microorganisms have intimate symbiotic relationships with other larger organisms; some of which are mutually beneficial (mutualism), while others can be damaging to the host organism (parasitism). If microorganisms can cause disease in a host they are known as pathogens.

Microorganisms are used in brewing, baking and other food-making processes. They are also essential tools in biotechnology and the study of biochemistry, genetics and molecular biology. They can also be harmful as a significant cause of human disease, and some have uses as biological weapons. Microorganisms have an important place in all ecosystems and in most higher-order multicellular organisms (as symbionts). They are vital to the environment, as they participate in the Earth's element cycles (such as the carbon cycle and nitrogen cycle). They are also involved in the recycling of other organisms' dead remains (see decomposition) and waste products.

Microbiology is the study of microorganisms, which are unicellular or cell-cluster microscopic organisms. This includes eukaryotes (with a nucleus) such as fungi and protists, and prokaryotes (without a nucleus) such as bacteria, protozoa and viruses (though viruses are not strictly classed as living organisms).

Although much is now known in the field of microbiology, advances are being made regularly. In actual fact, the most common estimates suggest that we have studied only about 1% of all of the microbes in any given environment. Thus, despite the fact that over three hundred years have passed since the discovery of microbes, the field of microbiology is clearly in its infancy relative to other biological disciplines such as zoology, botany or even entomology.

History of Microbiology

Bacteria were first observed by Anton van Leeuwenhoek in 1676 using a single-lens microscope of his own design. While Antony van Leeuwenhoek is often cited as the first microbiologist, the first recorded microbiological observation, that of the fruiting bodies of molds, was made earlier in 1665 by Robert Hooke.

The field of bacteriology (later a subdiscipline of microbiology) is generally considered to have been founded by Ferdinand Cohn (1828-1898), a botanist whose studies on algae and photosynthetic bacteria led him to describe several bacteria including Bacillus and Beggiatoa. Ferdinand Cohn was also the first to formulate a scheme for the taxonomic classification of bacteria.

Louis Pasteur (1822-1895) and Robert Koch (1843-1910) were contemporaries of Cohn's and are often considered to be the founders of medical microbiology. Louis Pasteur is most famous for his series of experiments designed to disprove the then widely held theory of spontaneous generation, thereby solidifying microbiology's identity as a biological science. Pasteur also designed methods for food preservation (pasteurization) and vaccines against several diseases such as anthrax, fowl cholera and rabies.

Robert Koch is best known for his contributions to the germ theory of disease, proving that specific diseases were caused specific pathogenic microorganisms by the application of what has become known as the Koch's postulates. Koch was one of the first scientists to focus on the isolation of bacteria in pure culture resulting in his description of several novel bacteria including Mycobacterium tuberculosis, the causative agent of tuberculosis.

While Louis Pasteur and Robert Koch are often considered the founders of microbiology, their work did not accurately reflect the true diversity of the microbial world because of their exclusive focus on microorganisms having medical relevance. It was not until the work of Martinus Beijerinck (1851-1931) and Sergei Winogradsky (1856-1953), the founders of general microbiology (an older term encompassing aspects of microbial physiology, diversity and ecology), that the true breadth of microbiology was revealed.

Martinus Beijerinck made two major contributions to microbiology: the discovery of viruses and the development of enrichment culture techniques. While his work on the Tobacco Mosaic Virus established the basic principles of virology, it was his development of enrichment culturing that had the most immediate impact on microbiology by allowing for the cultivation of a wide range of microbes with wildly different physiologies. Sergei Winogradsky was the first to develop the concept of chemolithotrophy and to thereby reveal the essential role played by microorganisms in geochemical processes. He was responsible for the first isolation and description of both nitrifying and nitrogen-fixing bacteria.

Types of Microbiology

The field of microbiology can be generally divided into several subdisciplines:

- *Microbial physiology:* The study of how the microbial cell functions biochemically. Includes the study of microbial growth, microbial metabolism and microbial cell structure.
- *Microbial genetics:* The study of how genes are organised and regulated in microbes in relation to their cellular functions. Closely related to the field of molecular biology.

- *Medical microbiology:* The study of the role of microbes in human illness. Includes the study of microbial – pathogenesis and epidemiology and is related to the study of disease pathology and immunology.
- *Veterinary microbiology*: The study of the role in microbes in veterinary medicine.
- *Environmental microbiology:* The study of the function and diversity of microbes in their natural environments. Includes the study of microbial ecology, microbially-mediated nutrient cycling, geomicrobiology, microbial diversity and bioremediation.
- *Evolutionary microbiology*: The study of the evolution of microbes. Includes the study of bacterial systematics and taxonomy.
- *Industrial microbiology*: The exploitation of microbes for use in industrial processes. Examples include – industrial fermentation and wastewater treatment. Closely linked to the biotechnology industry. This field also includes brewing, an important application of microbiology.
- *Aeromicrobiology:* The study of Airborne Microorganisms.
- *Food Microbiology:* The study of Microorganisms causing Food Spoilage.

BENEFITS OF MICROBIOLOGY

While microbes are often viewed negatively due to their association with many human illnesses, microbes are also responsible for many beneficial processes such as industrial fermentation (e.g. the production of alcohol and dairy products), antibiotic production and as vehicles for cloning in higher organsisms such as plants. Scientists have also exploited their knowledge of microbes to produce biotechnologically important enzymes such as Taq polymerase, reporter genes for use in other genetic systems and novel molecular biology techniques such as the yeast two-hybrid system.

Environmental Microbiology

At the turn of the 20th century, G. A. Nadson, a Russian microbiologist, proposed, microbes could serve as geologic agents, which control geochemical reactions. He astutely observed the role of microbes in the recycling, fixation and dispersion of elements in nature. The renewed interest to evaluate the impact of the microbiai factor in natural systems has been strongly influenced by the application of technical advances, which enable geochemists to image and measure microbial interactions and processes at various scales. Microbial activity has undoubtedly been modifying the Earth's uppermost lithosphere, hydrosphere and atmosphere since the appearance of life more than 3.5 billion years.

The ability of microorganisms to mediate the transformation of dissolved and gaseous substances and concentrate them in inorganic minerals, such as carbonates, pyrite or phosphates, provides evidence of microbial activity throughout geologic time. Additionally, this activity leads to the isotopic fractionation of elements, such as carbon, sulfur and nitrogen, providing tracers to interpret biogeochemical processes related to the evolution of life in Earth history. Thus, we have geochemical and mineralogical tools to interpret the geologic record, but, in order to construct convincing models, we need to validate the actual microbial processes through the study of modern environments and laboratory experiments. Indeed, the present holds the key to the past!

Knowledge about early microbial processes can, in fact, be obtained from the study of their modern counterparts. For example, studies of modern microbial mats and stromatolites provide crucial information to develop models for their formation in the geologic record. From the study of modern analogue environments, it is possible to combine geochemical and biological analyses to estimate the impact microbial activity may have had on paleo-environments. Molecular biology techniques indicate the presence of microbial consortium and distinguish among groups of microbes. Interactions within these microbial communities result in biogeochemical reactions, which generate specific bio-products, including gases, dissolved species, organic matter and biominerals. These products represent microbial "fossils", which are indicative of specific environmental conditions.

Modern environmental settings with peculiar physico-chemical, biological and mineralogical characteristics can be considered as analogues for "fossil" environments, which are defined as those that were common in the geologic past but are rare today. Assuming that the microbial processes have remained relatively constant under specific conditions throughout geologic time, a combination of geochemical and microbiologic studies in such "fossil" environments can provide a better understanding of the boundary conditions associated with the large-scale microbial biomineralisation observed in the geologic record. Further, the boundary conditions determined through the study of "fossil "environments can be used to design laboratory experiments to simulate the production of microbial "fossils "under controlled conditions.

By combining studies of modern analogues with well-defined experiments and observations from the rock record, realistic models to evaluate ancient geomicrobiological processes can be developed. For example, the origin of dolomite, a long-standing enigma in geology, is a geological problem with a probable geomicrobiological solution. In 1928, Nadson proposed that bacterial reduction of sulfate contributes to the precipitation of dolomite. He wrote, "Understanding the essential role played by this bacterial phenomenon may be the solution to the Dolomite Problem and the problems of the Mg cycle in the Ocean."

Recently, studies have been conducted in two hypersaline coastal lagoons located east of Rio de Janeiro, Brazil, which represent "fossil" environments in which modern dolomite precipitates. Based on field observations and measurements, culture experiments have been designed to simulate the natural environment, and microbial dolomite has been precipitated *in vitro*. Reproducing in the laboratory the processes and products observed in the studied "fossil " environments offers one of the best methods to validate the occurrence of microbial "fossils ", such as the biomineral dolomite. The study of microbial "fossils" produced in "fossil "environments combined with simulation experiments under controlled conditions provides geochemists a powerful tool to calibrate processes related to microbial activity in ancient systems. This geomicrobiological approach offers the possibility to glimpse into the environmental conditions of the early Earth and better interpret its evolving life forms.

Microbial Activities in Aquatic Environment

Evidence indicates that life originated in the primitive ocean approximately 3.5 billion years ago. The blood in our veins closely approximates the composition of sea water. Some biologists even consider terrestrial life an extension of the ocean that has developed the means of packaging sea-water so that it can navigate on dry land or in fresh water. Water has shaped evolution so that biochemistry must be considered from the perspective of its relationship to water. The concepts of the hydrophilic and hydrophobic nature of biological molecules; emphasising that they are always described in terms of this relationship to water. These water-loving/hating interactions ultimately determine the shape of biological molecules and in so doing, decide the *specificity* that is a universal principle of living processes. You have seen how the change in a single base pair in *DNA* that results in a subtle alteration in the shape of a crucial molecule can produce a genetic disease. Such alterations usually reflect minuscule interactions between a molecule's components and water.

All microbes live in an aqueous environment. Without water they either perish or become inactive. The ecology of aquatic environments is complex and our understanding of them is limited. Most aquatic environments are teaming with life. Microbes have evolved that can live in saturated salt (sodium chloride) solutions at temperatures from below freezing to >110°C; they thrive in waters full of toxic substances like copper, cyanide, lead, silver, gasoline, oil, benzene, and a plethora of other noxious natural and man-made substances. It is a *near natural law* that where liquid water, an energy source and basic nutrients exist, you will find living microbes. This is why the scientific community is so interested in finding water on Mars or any other planet/moon in our solar system.

Water is of vital importance to man for several reasons. First, as with all other life forms, it is required to maintain human existence. A person may survive for weeks without food, but only a few days without water. Water is crucial for the growing of our food and the operation of all our industries. Finally, it is a major vehicle of a number of water borne diseases that have shaped history in the past and will certainly continue to do so in the future. For example, the Middle East, where we recently went to war, is mostly a desert with a rapidly growing population of diverse people who really don't like one another. The Jordan River, the major source of water for Israel, Jordan and Syria, is, by standards here in the water rich Northwest, a trickle that we'd laughingly call a "river". Yet, the US is willing to defend one or more of the countries in this region, when their survival depends on this tiny tricking of water that's not much larger than the South Fork of the Palouse that runs through Pullman. Many people consider it likely that the next war in the Middle East may be over water not oil.

The importance of water in our lives is often overlooked because of its seemingly endless supply. Here in the NW water literally rains down upon us much of the year and if we drive a few miles in any direction we cross humongous rivers, plugged by tremendous damns and spanned by impressive bridges, that they tell us once contained a plethora of fish. Further, a half day's trip takes us to the Pacific Ocean, whose immensity and water content boggles the mind. Finally, 70% of the earth is covered with water, which is a whole lot of WET by anyone's definition. Surrounded by this plentiful supply of moisture it is hard to imagine that we could ever be short of water. The three descriptive terms in front of the word water changes the entire picture. These words are: Potable, Polluted and FRESH. We will be considering water from the microbial perspective of disease, but really we're talking about pollution and when we're discussing water pollution we're really talking about poop, feces, excrement etc. Time, unfortunately does not permit the examination of other related areas such as the diversity of microbial forms in aquatic habitats and their influence on the ecosystem.

Because of a need for a stable water supply humans always built their communities on or near a water supply. However, this has also led to an unfortunate tendency to dispose of our wastes in the same place where we obtain our drinking water. This presented little problem when the population was small and nomadic, for when the stink of a camp became offensive, our nomadic ancestors picked up and moved the tribe a few miles up or down stream or over the hill to the next stream. Mother nature then stepped in and rapidly recycled the few wastes left behind. With the coming of Agriculture, permanent settlements and large populations, it became impossible to move people to a clean habitat; besides someone usually was already there and they generally took serious umbrage when the new arrivals told them to move on. This probably accounts for the bad attitude of the American Indians towards the settlers.

Development of Sewage Treatment Technology

The first recorded toilets and sewers were built around 3,000 BC in several parts of the world, including the Orkney Islands and India/Pakistan. A system of channels on the Orkney Islands drained wastes from toilets in the homes into the sea, whereas cities in India/ Pakistan built open drains 2 ft deep by 7-10 inches wide along major streets through which the wastes were flushed away; many of these open sewers are still in use today. Cities in Iran/ Iraq had communal toilets in 2,300 BC and the Minoans of Crete had toilets and elaborate sewage systems in their palaces about the same time. Athens in 400 BC piped the wastes from homes to cesspools in the streets.

As the population and towns grew in size, the problem of obtaining fresh, clean water also increased. The Roman world is festooned with aqueduct ruins, some of which still work, at least over short distances. A major consideration of every Roman leader was insuring that the citizens of Rome had a constant clean supply of water. Water was brought to Rome from the mountains via an extensive system of aqueducts that spanned mountains and valleys. There were over 250 miles of water conduits inside Rome proper feeding a system of fountains that supplied citizens with fresh water. These fountains are still scattered throughout Rome today. Slaves, labored constantly repairing and maintaining the water system. In India and Pakistan remains of water pipes of clay sealed with asphalt dating back to 2,700 BC have been found. Copper water pipes were used in Egypt as early as 2,450 BC.

Every water supply system is constantly in danger of contamination. Before the Germ Theory of Disease few people recognised, other than from an aesthetic standpoint, the need for a "clean" water supply. You have seen how, by using statistics, John Snow, in 1855, connected polluted water with human disease. We realise today that throughout history most humans routinely drank contaminated water and that this situation still applies to the majority of humans on the planet. When the population was small and scattered, the outbreaks of water borne diseases were confined to concentrations of humans such as those in a single city or army.

However, as population increased and travel accelerated, the combination of improved disease transmission, crowding, poverty and ignorance resulted in extensive water borne epidemics and in the water being increasingly contaminated with human fecal waste containing pathogens. Microbes have evolved to take advantage of water as a mode of transmission of intestinal disease. Following the collapse of the Roman empire vestiges of the Roman water technology continued being used throughout Europe. Many villages and towns piped water to central fountains from nearby streams or springs, or diverted streams through channels in the towns. Centrally placed wells also supplied water in many cities.

Sewage disposal continued to be a severe growth-problem as the sise of cities grew. Many people, lacking toilets, used chamber pots which they emptied every AM in the streets or alleys. The stench in London in the summers throughout the middle ages and beyond was reported to cause grown men to faint and horses to collapse in their harnesses. Slowly however, the idea of sanitation began to take hold in the 1700s and accelerated through the 1800s. However, Queen Victoria's husband is reported to have died in 1861 of typhoid, indicating that even the Royal family apparently drank fecal polluted water. The idea of supplying water to the inhabitants of cities developed along with iron technology that made the manufacturing of inexpensive pipes possible.

Private water companies were formed throughout Europe to supply communities. The idea of filtering polluted water through sand to purify it appeared first for private homes of the wealthy and later for municipalities. With the establishment of the relationship between bacterial pathogens and water polluted with human wastes in the late 1800s, and worldwide communication that spread the stories of epidemics, people began demanding safe water; even if it involved taxation. To meet these demands it, in turn, became necessary to do something about the prodigious quantities of *raw sewage* spewing forth from cities and industries into the local water sources. As a consequence, sewage treatment technology developed in the early 1900s.

When a microbiologist inoculates a flask of bacterial medium with a culture the intent is usually to produce more bacteria. However, it can be viewed in another way.

The purpose could be to "use up the nutrients in the medium" by metabolising them. During the growth of microbes in a medium, the nutrients are oxidised to produce energy and new microbes. The result of this process converts most of the nutrients to chemicals like carbon dioxide, nitrate, sulfate, phosphate; i.e., minerals. Raw sewage is rich in organic nutrients such as human excrement, and food and industrial wastes. The purpose of a sewage treatment plant is to provide the optimum conditions for selecting and utilising microbes to convert easily biodegradable organic matter (EBOM) to mineral forms like carbon dioxide, nitrate, sulfate, and phosphate. This process is also called mineralisation. Since microbes grow and utilise nutrients most efficiently under aerobic conditions, sewage treatment plants are designed to provide excess oxygen for the microbes.

There is always some material that can not be easily degraded by microbes which *settles out* at various stages in the treatment process. This material is called sludge and it must also be disposed of as part of the sewage treatment process. There are two major designs of secondary sewage treatment plants (STP). These are called the activated sludge and the trickling filter systems. Each of them achieve the same efficiency of treatment, but they use different designs for doing so. The steps in sewage treatment are described below:

– *Step I*: The first two stages in both designs are the same. The first stage is the preliminary treatment. In this treatment the raw sewage coming out of the end of the sewage line runs into a large tank. While in this tank the sewage flow slows down enough so that heavy materials like sand and rock settle out and it passes through bars or screens which collect large floating debris like cloth, sticks etc. The effluent flows into the primary settling tank.

- *Step II*: primary settling tank is a large tank where the flow of sewage slows down to the point where largere pieces of organic matter settle out as sludge. These tanks also contain a skimmer that picks up smaller floating material, the majority of which are used condoms.
- *Step III*: The third step in the process involves the aerobic metabolism of the *EBOM*. In the trickling filter design this is achieved by spraying the effluent from the primary settling tank over a 4 to 6 foot pile of large inert blocks of material such as rocks, cement blocks, redwood blocks etc. The spraying saturates the effluent sewage with oxygen and air circulates between the blocks. The inert blocks become coated with a layer of microbes that are selected because they grow well in the sewage.
- *Step IV*: The tank's design is similar to that of the primary settling tank. Large debris, consisting mostly of chunks of microbial growth that flakes off the inert blocks, settles out in this tank as sludge. The effluent next flows into the chlorination basin.
- *Step V*: Sufficient chlorine is added to the effluent of the secondary settling tank to kill most the common bacterial pathogens that have survived the previous treatments. The treated effluent is then released into the environment, usually a nearby stream, river, lake or the ocean. This treatment however, does not kill (Sterilise) all of the pathogens present so it is unwise to frolic in the water around the sewage outfall.

In many areas , tap water is not considered safe to drink because it is polluted. That is, it is not potable water. In these places, usually large cities, the problem is often that the community has not been able, or willing, to build needed water purification facilities because of a lack of tax money. What frequently happens is that as the water supply upon which a community is dependent becomes more polluted, while the thirsty population has increased. This increases the demand for potable water, however current water purification facilities are unable to meet the demand and the citizens have decided, through their representatives, that increasing the capacity of the water purification plants is not a high priority.

To meet the added demand for water, the rate of throughput is increased. That is, the amount of water "purified" is increased by passing the polluted water through the purification system faster. When this is done, the purification process becomes inefficient and the water is not purified. Often the plant's operators simply increase the amount of chlorine added to the final effluent in an attempt to kill any pathogens that may have passed through the modified treatment process. However, since they are working at the "edge" of safety, they recommend that citizens should only drink "pure bottled water".

Because bottled water is inconvenient and expensive many people install "home purification systems" to remove pollutants. I'm unable to go into a detailed description of the design of these products as there are too many of them and more are appearing all the time as the problem grows. However, the bulk of these units involve one or both of the following components: a *filtration system* to remove microbes and other particles from the tap water and/or ion exchange resins or activated charcoal to remove the soluble dangerous minerals and/ or organic compounds. The ion exchange resins are the same things used in water softening tanks. Activated charcoal is a form of carbon that has the characteristic of absorbing large quantities of organic matter. The filters filter out particles based on the size of the holes in the filters.

Medical Microbiology

Medical microbiology is the study of parasites, fungi, bacteria, and viruses that are the agents of infectious disease in humans. Modern medicine relies on the control of microorganisms to maintain human health and quality of life. The divisions of medical microbiology include bacteriology, the study of bacteria that inhabit and/or colonize the human body and cause disease; mycology, the study of fungi as causative agents of human disease; parasitology, the formal study of the human parasitic organisms (protozoans, helminths, nematodes, trematodes and arthropods); and virology, the study of viruses that cause infectious syndromes in humans. Sizes for the pathogens considered include the smallest, viruses (50-100nm), bacteria that range from 0.1 μm (Chlamydiae) to 10μm (Bacillus rods), fungi ranging from ~8μm (yeasts) up to 10mm in size (filamentous fungi) and metazoan parasites that are visible to the naked eye.

Medical microbiology as a discipline requires a working knowledge of human anatomy and histology, and a comprehension of the pathologies associated with the infectious disease process. The human immune response to pathogens is key to the consideration of infectious disease. Understanding the relationship between pathogens and antimicrobial pharmacology is essential as well. Microbiology places information about pathogenic organisms and their specific characteristics within the context of host disease. Developing connections between microbiology and immunology will make learning more effective in both disciplines. Special patient populations are important to consider and should be appreciated for the distinctive infectious disease presentations they reflect. Age-related immuno-compromised status makes the elderly and the very young, especially newborn infants, particularly at-risk for respiratory infections. Health-related immunosuppression can predispose organ transplant recipients, and patients with immunodeficiency disorders, cancer, and diabetes to difficult infections. And, unique exposures due to occupation or travel can be a problem for some patient populations.

An immunocompromised person is lacking in some aspect of innate or adaptive immunity due to either a primary or secondary immune deficiency. Whatever the route, the net result is that the immunocompromised individual becomes susceptible to infection with a range of opportunistic pathogens from the commensal microflora and conventional infectious agents that cause a more severe form of disease than in a 'normal' host. This depends on how the patient's condition affects each limb of the immune system, which in turn controls the most likely pathogens. Immunity to infection and sensitivity to normal commensal microbes as pathogens varies throughout life and not just in 'disease' states.

Host-Parasite Relationships

A parasite is an organism that makes its living at the expense of another organism, so in a broad sense, all human pathogens fall under the blanket term "parasite". A parasite derives its nutrients and ecosystem from the host and is able to grow and reproduce in this environment. Medical microbiology presents concepts relating host-parasite relationships and the human immune system's response to infectious disease. Microbial ecology and system imbalance are keys to the understanding of many common diseases. The development of an infection and its eradication frequently involves issues that are defined as much by the host as by the microbial parasite, a balance between host immunity and the virulence of the infectious agent.

Normal microbial flora

The normal microbial flora of the human body is located mainly in the superficial layers and gastrointestinal tract. Gastrointestinal pathogens have some resistance to gastric acid and bile and agents of skin infections are resistant to drying. Lower respiratory and upper urogenital tracts are sterile normally, but they are susceptible to microbial "invasion" from adjacent sites.

Contamination occurs when microbes come into contact with host surfaces.

Colonisation. The skin of a newborn infant is populated initially by the organisms carried in the urogenital tract of the mother and acquired by the baby at birth. The mother's skin and breath are additional sources of normal bacteria that become the infant's commensal population, especially the mucosal normal flora.

The Oral Cavity. Many of the "normal" flora can be disease-causing organisms under the right conditions. Viridans streptococci are largely responsible for dental caries and pitting of tooth enamel by Streptococcus mutans acidic metabolic by-products is well documented.

Beneficial Effects. Vitamin synthesis and bile pigment degradation are two beneficial effects provided by bacteria to the host. Resident microflora inhibit the of growth of potential pathogens, as well as providing mechanical advantages such as suppression of competitor's adherence, antigen priming of the immune system, maintenance of low redox potential, and bacteriocin secretion.

Disease progression

The period of total infection is from point of first contact until point of complete elimination of the pathogen. Overt disease is associated with the onset of symptoms. In a recurrent infection, the disease reappears over time with a characteristic rise and fall of agent shedding and symptomatic disease. The seropositive phase occurs where the patient has been exposed to a pathogen and has mounted an antibody response that is reflected in detectable levels of specific serum immunoglobulin. Incubation is defined as the period from the point of first contact until the point of appearance of symptoms of infection (when "disease is present").

Latent infections remain dormant, and then re-emerge after stress or a lapse in immune function. During the period of communicability the infectious agent is being continuously shed by the patient. To complete the cycle of infection, infectious agents are excreted, and the route of excretion dictates the mechanism of spread. Fecal-oral spread involves excretion within stool samples and may be aided by diarrhea. Pathogens are often aerosolised in respiratory secretions by sneezing and coughing. STD agents are transmitted by sexual contact with vaginal, cervical or urethral fluids.

Zoonotic infections are diverse and cause diseases where humans are either a part of the normal infectious cycle or accidental hosts by contact with vectors or reservoir animals. Some of these agents are excreted in feces and urine, but also through parasitemia (widespread presence of the parasite). Examples include bloodstream parasitemia, to ensure uptake by blood-sucking insects (e.g. Anopheles mosquitos and malaria) and rabies virus budding from salivary gland cells to aid viral spread through animal bites.

Pathogen Virulence Mechanisms. Host-parasite interactions can be described as pathogenic vs nonpathogenic (pathogenicity=ability to cause disease or multiply within a host), or invasive as opposed to a successful commensal avirulent status quo (the human as an ecological niche). Virulence is a concept of "degree of pathogenicity". Some viruses and other pathogenic microbes characteristically cause disease and these may express highly specific adhesins or toxins.

Virulent organisms cause a specific host injury that is removed when the gene encoding a specific toxin or trait is deleted. The genes controlling the synthesis of these factors are grouped together under the control of a single promoter in pathogenicity islands. Examples of virulence factors include the neurotoxin secreted by Clostridium tetani (tetanospasmin) and the capsule of Streptococcus pneumoniae that subverts phagocytosis by preventing ingestion.

Toxin production is a common virulence factor and bacterial exotoxins can inhibit host cell protein synthesis, stimulate watery secretions, and cause violent symptoms of disease in the host. Attachment and adherence via specific epithelial and mucosal receptors (integrins) is another form of virulence because adhesion protects pathogens from the flushing of mucosal surfaces. Vibrio cholera binds to sites on the villi of the jejunum or ileum. Virulence is mediated by strength of attachment phenomena. Specifically pathogenic strains of Escherichia coli have capsular antigens (K1) that make the bacteria neurotrophic; most strains of E. coli are harmless commensal organisms that occupy space in the human gastrointestinal system.

Inhibition of host immune mediators. Inhibiting host immune cell products to blunt the anti-microbial response is a strategy of many pathogens. Some organisms produce leukocidins that kill neutrophils and macrophages. Others have intracellular growth patterns that allow them to avoid detection by the host's immune system (these will be discussed with the pathogen groups). Streptococcus pyogenes and Pseudomonas aeruginosa degrade human C5a using a peptidase enzyme. IgA protease is an immunoglobulin-destroying enzyme produced by Streptococcus pneumoniae, Neisseria meningitidis and Haemophilus influenzae. Complement compoenent C3b binding is inhibited by HSV envelope glycoprotein and vaccinia virus accelerates the decay of complement C3 convertase. Staphylococcus aureus produces catalase and this enzyme breaks down hydrogen peroxide in neutrophil phagolysosomes to limit bactericidal activity. Each of these mechanisms is a unique process for pathogen survival.

Molecular mimicry. Another pathogen survival mechanism is to change its surface composition, utilising a mask to avoid immune detection. One way to do this is simply to become coated with human "self-antigens". This method is used by Treponema pallidum (coats itself with fibronectin); Neisseria meningitidis (coats itself with circulating host IgA); and vaccinia virus (incorporates host sialic acid into its capsular polysaccharides). Presenting a changing target is the strategy for Trypanosomes and influenza virus (gene switching changes surface proteins).

Koch's postulates are an important set of criteria that can be used to judge whether a microbe is the cause of a disease in question. First, the organism must occur in every case of the disease, and must be responsible for the pathological changes observed. Next, the organism must cause the same disease in a new host exposed to an isolate, and must in turn be isolated from the new host.

Epidemiology

Epidemiology is the study of disease patterns and trends, of the occurrence, distribution and control of disease in populations. It also deals with disease tracking and prevention. Disease transmission is the movement of the infectious agent from one host to another. The risk of infection is dependent not just upon an individual patient's susceptibility, but on the level of disease within the population, the extent of population mixing and "herd immunity", added to the specific features of disease spread, such as communicable period, route and ease of transmission (infectiousness).

Modes of Transmission

Numerous modes of transmission contribute to the spread of human disease. For an infectious agent to persist within a population a cycle of transmission must be established leading from a contaminated source to a susceptible host and further propagating through the population.

Direct infection is the movement of a pathogen from human to human. Examples of organisms causing direct infection are Salmonella typhi and Shigella species, both of which infect humans only. Indirect transmission involves human to soil, water or surface dissemination of pathogens and infection of another human through contact with contaminated materials. This idea encompasses vehicle-borne transmission, whether through infected everyday objects or surgical instruments. Vibrio cholera is an example of a pathogen that is transmitted through contaminated water, especially in brakish coastal waters. Zoonotic transmission occurs through exposure to a nonhuman animal source of infection. The arthropod-borne agents like Borrelia and Rickettsia are included here as well as Spirillum, Brucella and Bacillus anthracis, that are contracted through animal bites or contact with animal products. Elimination is a potential outcome for a healthy host, whereby the parasite is eradicated at the end of infection cycle.

Routes of Transmission. The direct route means physical contact between humans or between a human and an animal to cause disease. Portals of entry include the gastrointestinal tract, respiratory mucosa, genital mucosa, and direct inoculation through the skin. Mucous membranes are especially important (STDs are transmitted in this way). The airborne route or respiratory droplet transmission is very important for viral pathogens and respiratory tract infections (aerosols). Fomites are inanimate objects contaminated with microorganisms, like drinking cups, towels and computer keyboards. The waterborne route is an especially important for enteric disease and it's an important route for fecal-oral transmission (ingestion). Vectorborne transmission is critical for some viral (arbovirus) & zoonotic infections (arthropod borne parasites).

Morbidity/mortality

Morbidity is the number of cases (or the incidence) of a specific disease per unit population (cases per 100,000). Mortality is the number of deaths due to a specific disease per unit population (deaths per 100,000). To illustrate morbidity/mortality concepts, for the U.S., Chagas disease has a low morbidity and low mortality, because it is present at very low incidence (the disease is not endemic, ie, not a typical pathogen for the region). Where Chagas' disease is present, it results in low mortality because the disease has a chronic nature, and death is not an acute outcome. In contrast, viral influenza has a high morbidity and relatively low mortality; many people are infected every year, but a low percentage of those affected will die from the disease. Ebola virus on the other hand has a low morbidity and a high mortality in the U.S.; it is hardly found, but is nearly always lethal. Prevalence is defined as "the number of cases of infection per unit of population at a single point in time". Disease incidence refers to "the number of new cases of infection per unit of population over a specified period of time". Disease levels vary throughout the year for some types of infection (for example the rise in respiratory tract infections during the winter months) with others remaining at a steady state (like urinary tract infection). Some infections have periodic cycles measured over longer scale, for example a four-year cycle for Mycoplasma pneumonia.

Patterns of infectious diseases vary throughout the world, depending upon factors such as environment, weather, economic factors, nutrition, behavior and the local healthcare systems. It is essential that every clinician is aware of regional disease patterns for effective diagnosis, treatment and control of infection within their own practice of medicine. For serious communicable diseases in the US, a part of surveillance is the legal requirement for medical personnel to inform the local health department of every diagnosis of a notifiable disease.

Epidemiological data can be presented as age-related incidence; geographical representation (syphilis); social group affiliation and distribution over time (AIDS). Diphtheria is the first component of the DPT vaccine (Diphtheria, Pertussis, Tetanus). For people 55 and older, there is an increase in morbidity because the DPT vaccine was unavailable 50 years ago, and data represented by age group shows that almost everyone had the disease at one time. For the example of morbidity and geographical variance, syphilis is especially high in the rural, southern part of the US, while its incidence is lower nationally. This is in part due to historical problems with healthcare access in rural areas with high minority populations (socioeconomic factors).

Finally, AIDS is a unique example of epidemiology in many respects. Initially thought to be a disease of specific population segments (Haitians, hemophiliacs, heroin addicts, homosexuals), it is now recognised to have more complexity (social group demographic). It is also a group of diseases that reflect primary underlying immunodeficiency, and this is a recurring theme throughout the study of infectious disease. And finally, the appearance of AIDS is changing over time, with increases in the incidence among the female and pediatric populations in the US, as well as in the MSM population (the CDC definition of "men who have sex with men" encompasses the highest-risk homosexual and bisexual male population).

Resisting the Threats of Microbes

Thirteen individual factors -some reflecting the ways of nature, most of them reflecting our ways of life -account for new or enhanced microbial threats. Any of these factors alone can trigger problems, but their convergence creates especially high-risk environments where infectious diseases may readily emerge, or re-emerge, afflicting individuals and societies alike while posing particular challenges for the medical and public health communities that must face these situations at the front lines. It's conceivable, in fact, that in certain places microbial "perfect storms " could occur -convergences of several factors -and unlike meteorological perfect storms, the events would not be on the order of once-in-a-century, but frequent. Following are the individual factors emergence:

- *Microbial Adaptation and Change*: The tremendous evolutionary potential of microbes makes them adept at developing resistance to even the most potent drug therapies and complicates attempts at creating effective vaccines.
- *Human Vulnerability*: Susceptibility to infection can result when normal defence mechanisms are impaired by causes such as genetically inherited traits and

malnutrition. Susceptibility can also result from antimicrobial resistance induced by the promiscuous use of antibiotics.

- *Climate and Weather*: Climate can directly affect disease transmission through its impacts on the replication, movement, and evolution of microbes and vectors; climate can also operate indirectly through its effects on ecology and human behaviour.
- *Changing Ecosystems*: Altered environments have immense influence on the transmission of microbial agents, whether waterborne, airborne, foodborne, or vector-borne.
- *Economic Development and Land Use*: Commercial activities can have intended or unintended impacts on the environment. For example, new or previously unknown infectious diseases have emerged from the increased human contact with animal reservoirs that resulted from changing land-use patterns.
- *Human Demographics And Behaviour*: Infectious diseases can result from individuals ' activities that involve exposure to microbial pathogens or simply from the increased probability of infectious disease as populations grow and people come into closer contact.
- *Technology and Industry*: Advances in medical technologies, such as blood transfusions and organ transplants, have created new pathways for the spread of certain infections. Meanwhile, the use of antibiotics in food-product animals has heightened antimicrobial resistance.
- *International Travel and Commerce*: The rapid and virtually unrestricted transport of humans, animals, foods, and other goods can lead to the broad dissemination of pathogens and their vectors throughout the world.

- *Breakdown of Public Health Measures*: In many places, the lack of basics such as potable water or sanitation contributes to infectious diseases. But similar effects can also occur elsewhere from inadequate vaccine supplies, low immunisation rates, or a paucity of expertise -say, in vector control.
- *Poverty and Social Inequality*: Mortality from infectious diseases is closely correlated with global inequities in income. Economic trends affect not only the individuals at risk but also the structure and availability of public health institutions necessary to reduce risks.
- *War and Famine:* Displacement caused by war and the fairly consistent sequelae of malnutrition from famine can contribute significantly to the emergence and spread of infectious diseases.
- *Lack of Political Will:* It is not only the governments in the regions of highest disease prevalence that must commit themselves, but also the leaders of affluent regions that ultimately share the same global microbial landscape.
- *Intent To Harm*: The world today is vulnerable to the threat of deliberate biological attacks that can cause large numbers of deaths and widespread social disruption. The likelihood of such events, in fact, is high, and public health systems and health care providers must be prepared to address them.

6

Modern Food Technology

For centuries, humans have been selecting, sowing, and harvesting seeds that produce food products that will sustain them. Global food demand has increased the need for improved crops. Biotechnology offers the needed technology to produce higher crop yields, plants that are naturally protected from disease and insects, and potentially more nutritious and better tasting foods. Benefits also can be seen in the environment, where insect-protected biotech crops can reduce pesticide use. Insect-protected crops allow for less potential exposure of farmers and groundwater to chemical residues, while providing farmers with season-long pest control. Also by reducing the need for pest control, less time, effort, and resources are spent on the land, and topsoil is preserved.

For decades, bakers, brewers, vintners, and ranchers have been using biology to their advantage and modifying genes to make bread, beer, wine, and cheese. The processes they have used can be likened to those used in agricultural biotechnology. One example is the process used in making cheese. Before biotechnology, most of the enzyme rennet, used to make cheese, came from the lining of calves' stomachs. The use of biotechnology has enabled researchers to remove the specific gene that produces rennet and reproduce it in bacteria. This allows the production of rennet through a fermentation process, eliminating the need for extracts from calves' stomachs. Currently, nearly 50 percent of the rennet needed in cheese production is produced through this fermentation process.

Yeast used in bread making has also been improved through biotechnology, resulting in a yeast that can speed up the leavening process in breads and pastries. Researchers in the United Kingdom developed this yeast by simply rearranging and duplicating certain yeast genes. Ancient bread makers might have been able to accomplish the same feat, but not without years of experimentation.

Development of Genetically Modified Food

A genetically modified food is a food product derived in whole or part from a genetically modified organism (GMO) such as a crop plant, animal or microbe such as yeast. Genetically modified foods have been available since the 1990s. The principal ingredients of GM foods currently available are derived from genetically modified soybean, maize and canola. Some governments have a very strong mutual disagreement over the labelling and traceability requirements for GM food products. For example the European Union and Japan require labelling and traceability while regulatory agencies in the United States do not believe these requirements are necessary.

The origins of genetic engineering represent a series of sequential scientific advances from the Nobel prize-winning discovery of DNA to the production of the first recombinant E .coli bacteria. The first commercially grown genetically modified food crop was a tomato created by Calgene called the FlavrSavr. Calgene submitted it to the U.S. Food and Drug Administration (FDA) for assessment in 1992; the agency determined that the FlavrSavr was in fact a tomato, did not constitute a health hazard, and did not need special labeling. Calgene released it into the market in 1994, where it was welcomed by consumers who purchased the fruit at two to five times the price of standard tomatos. However, production problems, and competition from a conventionally bred Long-Shelf-Life (LSL) variety prevented the product from becoming profitable, and Calgene was bought by Monsanto in 1995. A variant of the FlavrSavr was used by Zeneca to produce tomato paste which was sold in Europe during the summer of 1996. Its labelling and pricing were designed as a marketing experiment which proved that, at the time, European consumers would accept genetically engineered foods. This attitude would be drastically changed after outbreaks of Mad cow disease weakened consumer trust in government regulators, and protesters rallied against the introduction of Monsanto's Roundup-Ready soybeans.

Transgenic crops are grown commercially or in field trials in over 40 countries and on 6 continents. In 2000, about 109.2 million acres (442,000 km^2) were planted with transgenic crops, the principal ones being herbicide- and insecticide-resistant soybeans, corn, cotton, and canola. Other crops grown commercially or field-tested are a sweet potato resistant to a US strain of a virus that affects one out of the more than 89 different varieties of sweet potato grown in Africa, rice with increased iron and vitamins such as golden rice, and a variety of plants able to survive extreme weather.

Between 1996 and 2001, the total surface area of land cultivated with GMOs had increased by a factor of 30, from 17,000 km^2 (4.2 million acres) to 520,000 km^2 (128 million acres). The value for 2002 was 145 million acres (587,000 km^2) and for 2003 was 167 million acres (676,000 km^2). Soybean crop represented 63% of total surface in 2001, maize 19%, cotton 13% and canola 5%. In 2004, the value was about 200 million acres (809,000 km^2) of which

2/3 were in the United States. Four countries represent 99% of total GM surface in 2001: United States (68%), Argentina (22%), Canada (6%) and China (3%). It is estimated that 70% of products on U.S. grocery shelves include GM ingredients. In particular, Bt corn is widely grown, as are soybeans genetically designed to tolerate glyphosate herbicides. The US Agriculture Department estimated that 38 percent of the 79 million acres (320,000 km^2) of corn planted in 2003 will be genetically engineered varieties as well as 80% of the 73.2 million acres (296,000 km^2) soybeans. The Grocery Manufacturers of America estimate that 75% of all processed foods in the U.S. contain a GM ingredient.

Future envisaged applications of GMOs include bananas that produce human vaccines against infectious diseases such as Hepatitis B, fish that mature more quickly, fruit and nut trees that yield years earlier, and plants that produce new plastics with unique properties. The next decade may see exponential progress in GM product development as researchers gain increasing access to genomic resources that are applicable to organisms beyond the scope of individual projects.

Crop Breeding

When crossing two plants using traditional breeding techniques, breeders mix thousands of genes in order to transfer the protein products to enhance one or a few genetic traits. Therefore, the odds of something—such as a naturally occurring toxin—being transferred unintentionally are far greater in traditional breeding than in biotechnology. To avoid toxicity, breeders spend many years back-crossing the new plant varieties with distant relatives. Back-crossing involves crossing new plant varieties repeatedly with plants whose genetic components are well known. This slowly dilutes the impact of all those unwanted genetic traits that came along with the few beneficial traits.

When breeders develop new varieties, they carefully consider two questions:

– Will they inadvertently increase naturally occurring toxicants while seeking other modifications, such as enhanced protein content?

– Might they inadvertently create a toxicant in a new plant by crossing two plants known to be safe?

Plant breeding is generally safe, and biotechnology can make it safer. Biotechnology brings to traditional plant breeding the ability to move single genes instead of having to move thousands and makes possible the identification of those genes and their protein products that are toxins. Rather than spending 10 to 12 years to breed plants in the traditional manner and mixing thousands of genes to improve a crop plant, modern crop breeders can select a specific genetic trait from any plant and move it into the genetic code of another plant through biotechnology.

As the knowledge of biotechnology grows, toxicologists can improve their understanding of how natural and synthetic toxicants actually affect humans and animals at the cellular level. This means that "high dose" animal tests, in which various substances are fed to research animals in high doses to assess possible human health risk, will be steadily replaced by "low dose" studies in which the effects of tested substances on cultures of animal cells can be studied directly. This also means greater accuracy and utility.

Benefits of Food Biotechnology

Farmers have relied for centuries on the newest technology to produce and enhance foods that possess specific beneficial traits. Biotechnology offers farmers a more precise way to accomplish these goals. The use of biotechnology benefits not only the grower, but also the consumer. Growers reap higher crop yields, while consumers have greater product choices year round.

Current benefits of biotechnology include:

- Disease resistance
- Reduced pesticide use
- More nutritious composition of foods
- Herbicide tolerance
- More rapid growth of crops
- Improvements in taste and quality.

Some of the products that have been enhanced through biotechnology are:

- Corn, soybeans, and cotton that require fewer applications of herbicides/pesticides
- Soybeans that are lower in saturated fats, and higher in oleic acid, offering better frying stability
- Virus-resistant papayas that make crop production more dependable
- Peppers improved to be tastier (sweeter) and remain firmer after harvest
- Potatoes and corn that are disease-resistant and produce a higher yield.

The volume of biotech crops in development continues to grow. Biotechnology has been used in a number of crops for several years, and more genetically enhanced products are expected to be on the market in the coming years. Benefits that can be expected in the near future include:

- Reducing levels of natural toxins, such as allergens, in plants
- Providing simpler and faster methods to locate pathogens, toxins, and contaminants
- Extending freshness.

Products that could be on the market in a few years as a result of these developing benefits include:

- Oils, such as soybean and canola oils, developed to contain more stearate, making margarine and shortenings more healthful
- Peas grown to remain sweeter and produce higher crop yields
- Smaller, seedless melons for use as single servings
- Bananas and pineapples with delayed ripening qualities
- Peanuts with improved protein balance
- Fungus-resistant bananas
- Tomatoes with a higher antioxidant (lycopene) content than current varieties
- Fruits and vegetables containing higher levels of vitamins such as C and E to potentially protect against the risk of chronic diseases such as cancer and heart disease
- Garlic cloves that produce more allicin, possibly helping to lower cholesterol levels
- Higher-protein rice, using genes transferred from pea plants
- Strawberries that contain increased levels of ellagic acid, a potential cancer fighting agent

- Peppers, strawberries, raspberries, bananas, sweet potatoes, and melons that are enhanced for better nutrition and quality
- Strawberries with higher crop yields and improved freshness, flavor, and texture.

According to the U.S. Census Bureau, the current world population is about six billion and projected to grow to about nine billion by 2050. Few other technologies will be able to approach biotechnology's potential to help avoid starvation in this century. By increasing a crop's ability to withstand environmental factors, growers will be able to farm in parts of the world currently unsuitable for crop production. Along with additional food, this could also provide the economies of developing nations with much-needed jobs and greater productivity. Biotechnology will also enable growers to produce further enhancements in plant varieties. This would allow for the possibility of increasing the agricultural gene pool that billions of people rely on for basic foodstuffs.

Another economic and environmental benefit is seen in the area of fertilizer use. Almost half of the $12 billion American farmers spend each year on fertilizer simply evaporates or washes away. As a result, much of the fertilizer used is wasted and can end up in water sources, damaging the environment. Some plants, such as corn, might be improved to draw nitrogen from the soil, thereby reducing the need for fertilizer.

Products we can expect to see further down the road include:

- Enhanced protein quality in corn and soybeans (increased levels of lysine and methionine for improved animal feed)

- Modification of acid production in potatoes, allowing potato plants to withstand frost
- Rice that grows faster, is richer in protein, and can better tolerate drought and acidic soils
- Rice enriched with vitamin A that may reduce incidence of blindness in developing countries

Even under the best of conditions, food production for hundreds of millions of Americans—and billions more around the globe—can take a toll on the environment. Erosion can claim precious topsoil; farm chemicals sometimes reach streams, rivers, and groundwater supplies; and livestock can deplete grazing lands. Wetlands and other sensitive habitats sometimes get plowed under for use as farmland. And in the world's tropical forests where an estimated 90 percent of the world's species exist, poor farmers clear trees in order to provide food and a living for their families. By improving many aspects of modern agriculture, biotechnology can help alleviate many of these pressures on the land, both by preserving natural resources and by reducing environmental stresses.

Many of today's fungicides, herbicides, insecticides, and other pesticides are better, safer, and more environmentally sensitive than older versions. Even so, they sometimes enter the air, soil, and groundwater when they blow or wash off plants. Biotechnology can achieve many of the goals for which pesticides were designed—often more efficiently.

Farmers recognize more than anyone that healthy growing environments define their future. Thus, they always seek better ways to control weeds with the least toxic herbicides available that do not damage food crops. Farmers also would like to reduce their use of insecticides and fungicides, limiting their own exposure to the chemicals. Growers have strong economic incentives to reduce agricultural inputs.

Erosion of topsoil due to wind and water can be cut by more than 70 percent—in some cases up to 98 percent—when farmers use no-till techniques, meaning they do not plow under weeds and crop residues after harvesting or before planting. Eager to protect their soil, American farmers have flocked to no-till farming. Acres committed to no-till reached nearly 53 million in 2000. Herbicide-tolerant canola, cotton, and soybeans allow improved weed control, and therefore have enabled greater use of soil conservation farming practices.

As much as 75 percent of American corn and soybean crops each year are eaten by farm animals. Anything that makes animals better able to digest their food will diminish the overall need for feed, without diminishing the animals' productivity.

Other feedstock-related benefits of biotechnology include:

- Improved protein quality by balancing amino acids, thereby reducing nitrogen in waste
- Reduced environmental impact of phytate in animal waste
- High-oil corn that results in high energy density, therefore resulting in more meat per ton of feed.

Even though the Earth's atmosphere contains about 78 percent nitrogen, most crops have no mechanism to use this natural nitrogen. As a result, farmers depend on added fertilizers to provide the nitrogen necessary to boost crop yields. But crops only use about 50 percent of the more than 60 million pounds of nitrogen fertilizers added to them each year. The excess nitrogen can cause environmental problems in soil and water. Growers have long recognized and used the innate abilities of legumes like soybeans to "fix" nitrogen, which means to use the natural nitrogen in the soil and air. These natural nitrogen fixers replenish the nitrogen supply in the soil from which they were harvested. Breeders desire to develop other crops that can "fix" their own nitrogen, putting such plants high on researchers lists.

Should breeders succeed in creating the "self-fixers," they would:

- Allow farmers to decrease their use of synthetic fertilizers while maintaining bountiful yields
- Result in less nitrogen from fertilizers remaining in the soil or running off into the water
- Greatly enhance productivity in many regions of the developing world where farmers cannot afford nitrogen fertilizers.

Corn and soybeans can become natural factories for production of ingredients like sucrose, lysine, and methionine that are used for animal feed. These crops would essentially be recyclable and biodegradable and replace industrial factories, thus reducing stress on the environment. Of the more than 80,000 species of edible plants known to exist, humans cultivate only about 300 of them. Of those, only about 12 have emerged as major staples. Through genetic enhancement, crop breeders can: Increase the use of plant species by using biotechnology to discover which genes of value reside in which plants and then transfer those genes into crops now in use around the globe.

- Expand the genetic variation in staple crops by breeding into them desirable traits from previously unavailable sources. This will not affect the relatively narrow genetic lineage of many crops in the near term. Longer term, it will significantly expand the gene pool used in modern agriculture and thus reduce the relatively low, but real, risk of crop failures.

– Expand many wild relatives of modern crop plants that might be threatened with extinction.
– Finally, enable scientists to learn what important genes are actually contained in the millions of plant specimens housed in gene banks around the world.

Risks of Food Biotechnology

The majority of commercially available crops have an agronomic advantage like herbicide tolerance or insect resistance. These traits offer major benefits to the farmer. However, there are indirect benefits to the consumer from these traits: GM crops have shown to contribute to significantly reduced greenhouse gas emissions from agricultural practices. This reduction results from decreased fuel use, about 1.8 billion liters in the past nine years, and additional soil carbon sequestration because of reduced ploughing or improved conservation tillage associated with biotech crops. In 2004, this reduction was equivalent to eliminating more than 10 billion kg of carbon dioxide from the atmosphere.

Controversies surrounding GM foods and crops commonly focus on human and environmental safety, labeling and consumer choice, intellectual property rights, ethics, food security, poverty reduction, and environmental conservation.

Dangerous Potato

Although no major health hazards have come to light since GM food was introduced 12 years ago, some fear for the long term health risks which GM could pose, or that the risks of GM have not yet been adequately investigated. In August 1998 widespread concern, especially in Europe, was sparked by remarks by nutrition researcher, Dr Arpad Pusztai, regarding some of his research into the safety of GM foods.

Pusztai claimed his experiments showed that rats fed on genetically modified potatoes had suffered serious damage to their immune systems and shown stunted growth. He was criticised by leading British politicians, the majority of scientific peers with expertise in the area and by the GM companies in a television interview, preceded the scientific publication of his results. When his studies were finally published in The Lancet, no evidence of stunted growth or damage to immunue system was substantiated. The Royal Society's review of the Pusztai data led to the conclusion that the study "is flawed in many aspects of design, execution, and analysis and that no conclusion should be drawn from it".

Alternate view of Pusztai, supported by the evidence available. The conclusion of the potato feeding study were attacked by the biotechs but showed much evidence of damage. The rats fed on the GE potatoes suffered substantial damage to their immune systems, plus abnormal weight loss in various organs, including the brain, testicles and liver. Some also had abnormal growth in their intestinal cells, which could indicate a prelude to cancer. Despite a long and concerted industry campaign to discredit Pusztai, part of the results of his GE potato study were duly peer-reviewed and published in the scientific literature. However, despite the industry and scientific establishment dismissing his work, there has been little attempt made to independently repeat and confirm the study, as is usually the case in science. Pusztai has published other results of his experiments since, which raise serious question over the safety of GM foods.

In the USA Food and Drug Administration regulates the safety GM food using the same standards and requirements as under U.S. food law generally. Substances added to food that do not meet the statutory definition of "generally recognised as safe," and that are not pesticides, are classified as food or color additives and must be pre-approved before they may be marketed.

Manufacturers GM foods are held responsible for the safety of their products under both general US law and general provisions of the Federal Food, Drug, and Cosmetic Act. In addition prior to marketing these GM foods manufacturers are required to submit to FDA documentation demonstrating their safety and await approval for their use.

The claim of manufacturers being held responsible flies in the face of every fact available. In May of 2006 Congressman Dennis Kucinich introduced 6 bills addressing the gaps in legislative protections in the GMO food category. Among them is H.R. 4816 The Genetically Engineered Organism Liability Act would hold the companies responsible, but this is far from the case at present time.

Some people raise concerns that the safety of these novel food forms relies on the voracity of the manufacturers doing the testing. Defenders of the regulatory model point out that such a model has served the public well for many years with respect to non-GM food, drugs and food additives. They further point out that manufacturers data is subject to close scrutiny and that the law protects the consumers from manufactures of dangerous foods. It is thus in the manufactures best interests to ensure that the safety testing of GM foods is done adequately. This reflects the opinion of the pro-industry AgBioWorld.

Many believe regulatory models fail to protect consumers and are subject to pressure and influence by industry. In addition many FDA approved drugs have been recalled in the wake of unreported side effects or supressed findings.

Dangerous corn

Another controversy recently arose around biotech company Monsanto's data on a 90-Day Rat Feeding Study on a strain of GM corn. In May 2005, critics of GM foods pointed to differences in kidney size and blood composition found in this study, suggesting that the observed differences called into question the regulatory doctrine of substantial equivalence - that GM food with similar proteins and toxins is deemed no different from conventional food, without further investigation of the effects of any other differences. Some argued that this study suggested human health might be affected by eating GM food.

However, the EU regulatory authorities that examined the Monsanto data concluded that the observed small numerical decrease in rat kidney weights were not biologically meaningful, and the weights were well within the normal range of kidney weights for control animals. There were no corresponding microscopic findings in the relevant organ systems, and all blood chemistry and organ weight values fell within the "normal range of historical control values" for rats. Thus, the experts concluded that there were no effects on the functioning of kidneys in rats fed a diet of GM corn.

New Allergens

GMOs that induce allergies have been produced in the laboratory. In 1993 Pioneer Hi-Bred International developed a soybean variety with an added gene from the

Brazil nut. This gene increased the levels of methionine, a nutrient commonly added to poultry feed, in the GM soybean. However, a preliminary Pioneer funded study by the University of Nebraska indicated that the added gene could cause allergic reactions in humans. The completed study, published in the New England Journal of Medicine, later confirmed the preliminary results. Pioneer discontinued further development of the GM soybean and had all material related to the modified soybeans destroyed. While this study indicates the possible risks of GM foods, some point out it establishes the commitment the developmental community has toward consumer safety as well as the competence of current safeguards. A similar result was published in November 2005, when a pest resistant field pea developed by the Australian CSIRO for use as a pasture crop was shown to cause an allergic reaction in mice. The immunologist who tested the pea noted that the episode illustrated the need for each new GM food to be very carefully evaluated for potential health effects.

Environmental and Dcological Impacts

There is some evidence for positive impacts of the planting of GM crops on reduced green house gas emmisions and pesticide loads in the environment. However, there has been controversy over the results of a farm-scale trial in the United Kingdom comparing the impact of GM crops and conventional crops on farmland biodiversity. Some claimed that the results showed that GM crops had a significant negative impact on wildlife.

Others pointed out that the studies showed that using herbicide resistant GM crops allowed better weed control and that under such conditions there were fewer weeds and fewer weed seeds. This result was then extrapolated to suggest that GM crops would have significant impact on the wildlife that might rely on farm weeds. In July 2005 the same British scientists showed that transfer of a herbicide-resistance gene from GM oilseed rape to a wild cousin, charlock, and wild turnips was possible.

As far back as 1996, a weed that could resist five times the recommended Roundup dosage was found in Australia, and in 2000 scientists discovered a herbicide-tolerant canola plant that cross-pollinated with a related weed. That same year, canola weeds resistant to three herbicides, were reported in western Canada. Since then, reports of glyphosate-tolerant weeds have only increased.

Has Roundup Ready soy been a good deal for the American farmer? In Illinois the use of RR soy has resulted in the most expensive weed control system in modern history: between $40 and $60 per hectare. Before the introduction of RR seed, such costs averaged $26 per hectare. In addition, RR soy yields are on average no larger than those of non-RR varieties, and in fact are sometimes lower. A 1998 Iowa farmer survey found RR soy to yield 4% less than conventional soy. And then there's the seed's cost. An Iowa farmer can spend $26.42 per acre on RR soy seeds, whereas conventional varieties cost $18.89 per acre.

Bt Crops

The industry claims that Bt crops have greatly reduced pesticide use. But whether less pesticide is actually used because of Bt crops is an open question. A 1999 study by the USDA's Economic Research Service showed no statistically significant difference in pesticide use between Bt and non-Bt crops. In fact, it found that in the Mississippi Delta, significantly more pesticides were sprayed on Bt crops. But the greatest problem is the development of pest resistance to the Bt toxin, warns UC Professor Miguel

Altieri, "No serious entomologist questions whether resistance will develop or not. The question is, how fast?"

In Makhathini Flats, South Africa, the majority of small-scale farmers that used Bt cotton have stopped planting it because they could not repay their debts. A five-year study by Biowatch South Africa showed most farmers that planted Bt cotton had not benefited. In India, Bt cotton failed huge numbers of farmers in Andhra Pradesh and Madhya Pradesh, many of whom were driven to suicide as a result of heavy debts from purchasing Bt cotton seed, which was 3-4 times the price of conventional cotton.

Bt crops can also harm beneficial insects and adversely affect soil ecology. Adverse effects of Bt crops on beneficial insects were known at least as far back as 1999, when research led by Charles Losey of Cornell University discovered that Bt corn pollen was toxic to monarch butterflies, under laboratory conditions. Losey came under withering attack by pro-industry scientists, as were Pusztai, Chapela and Quist in their moment, but Losey's critics ignore that subsequent research confirmed that Bt crops indeed are a hazard to "non-target" species.

Ethical Aspects of Food Biotechnology

The initial developments in food biotechnology, focused on the GM of micro-organisms to enhance production of enzymes for use in food manufacture. The application of GM in these areas, remains the most extensive use in food production to date. A further major area of application has been the development of a wide range of crops modified for agronomic traits. The main traits developed for commercial use to date have been: herbicide resistance and insect resistance (e.g. in maize, Soya bean, cotton, oil seed rape), and virus resistance (e.g. potato).

Experiments are increasing rapidly on crops modified for nutritional and health advantages in the final food product, so called 'functional foods' and 'nutriceuticals', both for human consumption and animal feed. In addition, crops and animals are being modified for production of pharmaceutical benefit, so called 'pharming'. The metaphor is used of crops becoming factories, producing: vaccines (e.g. the polio vaccine in a banana), plastics, industrial starches, and feed supplements and enzymes.

The initial developments are directed to the food production in the developed world. Promised modification of crops for growth in the difficult conditions in the developing countries are not at the marketing stage whereas, herbicide and insect resistant plants are being extensively planted in North America, Australia, China, and increasingly Latin America. The increased commercial planting of GM crops has in turn brought increased public awareness of the entry of GM into the food chain.

The identification of the important social and ethical aspects of food biotechnology for EU consumers can be derived from an analysis of both quantitative surveys, conducted by Eurobarometer, and qualitative surveys carried out in The Netherlands and the UK The Eurobarometer Survey 46.1 'The Europeans and modern biotechnology' , supplemented two earlier surveys from 1991 and 1993 on European citizen opinion on biotechnology. The findings of this survey include the following salient observations:

Knowledge and anticipated effects of the technology: There is a correlation amongst respondents between a greater objective knowledge and both optimistic and pessimistic views of its anticipated effects. Greater knowledge does not bring either increased support for biotechnology nor greater opposition.

Perceptions of benefit and risk of applications: Perception of risk is higher amongst those with greater objective knowledge and those who have discussed biotechnology over recent months, but such perception is low amongst those with little knowledge. Overall Europeans show awareness of both benefits and risks of Biotechnology—a mixed stance that has been mirrored in surveys in countries such as Canada and Australia.

Labelling: The majority of the respondents (74%) favour labelling of genetically modified foods.

Regulation: Overall, the findings suggests a lack of confidence in self regulation, but also a relative lack of trust and confidence in the effectiveness of EU and national regulation and institutions, suggesting worrying implications for the legitimacy of the governance of modern biotechnology.

Confidence and trust: The lack of confidence in government, is echoed in the response to the question as to the most reliable source of information on biotechnology. The responses were as follows:

- Consumer organisations.
- Environmental protection organisations.
- Schools: universities.
- Public authorities.
- Industry.

A problem for surveys of attitudes to biotechnology, has been the lack of prior knowledge and understanding of the technology amongst the public. However respondents can, and do, draw on their existing cognitive frameworks to formulate their attitudes on the technology and its applications. Hence, in addition to these quantitative surveys, more qualitative attempts have been made to assess consumer attitudes to GM food, by seeking to relate consumer's attitudes to their more deeply held concerns and values. These surveys were conducted in The Netherlands and the UK. The key values which emerged were based around perceptions of trust, choice, need and care for a sustainable society, which included conceptions of natural balance.

Lack of trust in the regulatory process and the credibility of regulations emerged from research in the UK and was linked to a weak sense of agency in the regulatory process . Conversely, in The Netherlands where the respondents felt trust in the regulatory process and authorities, then they also felt that they had control. However, this trust can be undermined by fear, which can be induced by lack of knowledge of food production and how to interpret information on the packaging.

Variation in products and variation in prices, as well as, consumer knowledge about the interpretation of information on packages of foodstuffs and the production methods. Because of a lack of knowledge and uncertainty over the effects on health and natural balance, biotechnology undermines this freedom of choice somewhat and engenders some fear and insecurity. The notion of choice was a complex issue. On consumer choice, the low price of a food was a key determinant, but was also seen as an economic constraint against real personal choice. Labelling was seen to be of limited utility because of the desire to have information about the social context of the decision to produce a food through GM. Choice as a citizen was also an important factor, but here there was resignation and, once again, a lack of a sense of personal agency.

Applications of food biotechnology that are seen as important, or needed, are supported, as opposed to those that are seen as trivial. Perceived need was associated with agricultural applications in which modification appeared to improve production as opposed to the end product. Consumers are at pains to discriminate between classes of product and the different issues, negative and positive, that these raised. For example, whether the use of herbicide was increased or reduced was seen as an important criterion for assessing GM crops.

The value of care for a sustainable society covers concerns over: the natural balance, the usefulness or necessity of the application of modern biotechnology, trust, health, social dissipation and third world problems. Natural balance covered recognition of potentially positive implications of biotechnology such as: possible cleaner, less chemical dependent, and more efficient food production protecting the environment and using fewer raw materials, and the preservation of rare species through cloning. Negative effects were also identified such as: decline in crop diversity, as super crops may come to predominate food production, overproduction, with threats to the environment and ecosystems, and irreversibility of harm, for example upon different trophic levels in the natural food chain.

Individual choice can also be seen from the perspective of the consumer, as well as the citizen. The notion of informed choice in the market place is the basis of the concept of consumer behaviour. This requires the availability of different types of products, adequate labelling and knowledge to allow informed choice. The nature of the citizen, in contrast, refers to the individual in relation to governance. A situation in which a majority of the Europeans reject or accept biotechnology: GM on ill-defined grounds is undesirable. Therefore, an extensive information campaign is needed to allow Europeans to make an informed choice. The information to be provided should consist of several elements, including:

- A basic understanding of GM and food production processes through the food chain.
- Advantages and disadvantages of GM as portrayed from different perspectives.
- A system of labelling, including the labelling of products.
- And finally, it must clarify the decision making process and involve a wide range of information providers.

However, the role of information is limited if there is no basic education in this field. Therefore, a good longer term strategy is to build the above elements in the European education system. This is currently the responsibility of member states, but the European Commission has a role to play. A major instrument for making informed choice has always been considered the labelling of food products.

The basis of the European regulation of GM food is that only food that is no longer equivalent to non GM food should be labelled, as laid down in EC Regulations: 258:97 and 1139:98. The method of determining equivalence, as laid down for the food products of GM Soya and maize in 1139:98, is through the detection of modified DNA or protein in the final food. The entry of GMOs into the food chain and the massive commingling of GM crops with non GM crops has clouded the efficacy of labelling the final product. This is especially the case for compound foodstuffs that contain ingredients, such as oils and lecithin, derived from raw materials that have undergone modification at an earlier stage of the food chain. Such ingredients may not be detectable.

Many retailed foods are likely to contain ingredients that originate from GM crops but are not labelled. Also, in large scale food industrial processes and commodity trading systems, contamination can not be avoided. This can only be accommodated by the application of threshold levels. In addition, the labelling system must be verifiable. For consumers, wishing to make their choice of food product based on the relatively sophisticated and differentiated bases, as outlined previously, such as the impact upon the environment and upon sustainability of the form of agricultural production, labelling will prove to be neither informative nor transparent. Although the labelling system is not entirely consistent, it is in itself a useful element of consumer information. A comprehensive review of the system to get more uniformity is desirable.

Complementary to labelling, transparency right through the food chain must be created, by the use of a comprehensive system of segregation and certification of GM crops and their products from non GM crops at each stage of the food chain, and for this to be reflected in the final labelling information. There is ample precedence for such a comprehensive system of transparency as in the case of halal meat production and organic food certification. This degree of transparency would allow consumers to make a more fully informed choice of foodstuffs, in line with their more deeply felt values on such issues, and would provide for a more democratic and participatory basis for transparency.

With the increased use of GM crops, segregation will become even more important. The public's need, and expressed desire, for information on food production and content need to be acknowledged and incorporated into regulatory decisions, so that consumers can express their preferences in the market place. In the case of food biotechnology such risk may become more acceptable where there is transparency and openness throughout the processes of production in the food chain and at the point of consumption. This is reflected by public support for the labelling of the products of modern biotechnology and full information about their food. Such transparency may need to include traceability throughout the food chain.

Greater public knowledge and access to information alone will not guarantee public acceptance of the products of food biotechnology. There is a need to reconnect the public to the decision making, governing the regulatory processes for food biotechnology, in order to build public trust and confidence in this decision making. This will involve decision making, which is seen to take the differing views of the public, both as consumers and as citizens with different social interests into account. In order to incorporate these views, greater understanding is needed about the basic values of the different publics throughout Europe and how these values inform their attitudes towards food biotechnology. These qualitative surveys need to embrace all of the member states in Europe. These surveys should be sponsored by the European Commission and member state governments. The findings of such surveys must be incorporated into the decision-making processes, starting at the level of R&D policies through to the risk analysis of GMO and food biotechnology products.

Further techniques will be utilised to incorporate the public's views, which may include the use of citizens' juries and consensus conferences. This wider inclusion into the decision making process should engender greater social acceptability and trust in the regulatory processes through greater public involvement and understanding. This involvement will lead to a more informed and effective regulatory process. Public acceptance may also be enhanced through extending this more open process of regulation into more informed consumer choice in the marketplace through greater transparency.

The development of modern biotechnology may contribute to the move towards a sustainable society if it meets with the following conditions, which take into account the concerns and values of consumers:

- The applications fit in to more ecologically balanced agro-systems, such as rapid detection methods for infections with pests, diseases or other types of stress in plants.
- The applications lead to less waste in crop production and diminish the need for chemical pesticides.
- The benefits can be justified to consumers, in terms of values that they perceive as important in food production, for example ensuring the application can be linked to a 'more natural' form of production.
- Awareness is shown in the development of the technology of the possible negative impacts that its deployment might have on the interests of small-scale farmers in the developing countries, for example the development should focus on low-tech, easily adaptable and cheap applications for developing countries.
- Greater trust is created: by ensuring transparency over the deployment of biotechnology throughout the food chain, from laboratory and field research, through commercial growing, commodity handling, and food processing, to retail and consumption of the products of food biotechnology.

Risk Regulation

Attempting to predict the effects of the use of GM in agriculture and food production has raised several scientific uncertainties. These are the focus of scientific research projects within Europe. Scientific opinion is divided about the importance of these uncertainties. These uncertainties include:

- Incomplete scientific knowledge about how genes work in a specific environmental context to be able to predict with certainty what will happen when genes are transferred from one organism to another, and about the stability of the transferred genes.
- The possibility of 'escape' of transgenes from their host organism into others.
- That the GMOs themselves will become invasive or persistent.
- Other unintentional and unpredicted interactions between GMOs and their environment.
- GMOs and food safety.
- Effects upon agricultural practice.
- Secondary impacts upon bio-diversity.

These scientific uncertainties are reflected in consumers' perception of the risks of the release of GMOs into the environment and the consumption of GM food.

An example of a template commonly used for the process of risk analysis for food and environmental safety is the scheme defined by the FAO: WHO Consultation on Risk Analysis held in Geneva in 1995. This scheme defines risk analysis as including:

- Risk assessment.
- Risk management.
- Risk communication.

Risk assessment consists of: (i) hazard identification, (ii) hazard characterisation, (iii) exposure assessment and (iv) risk characterisation. Risk management is the process of managing the output of the risk assessment step (the risk characterisation). Risk communication is a two-way exchange of views covering the whole process. The European Commission proposed that the process of risk assessment of food safety should be seen as a scientific process, to be conducted by scientific advisors. The processes of risk management and communication were identified as the political processes. The basis for risk assessment is scientific, however the experience with GMOs and GM Food is limited. These uncertainties have led to deployment of the precautionary principle to the process of risk analysis of GMOs.

Precautionary Principle

The precautionary principle is an approach to risk management that is applied in circumstances of scientific uncertainty reflecting the need to take action in the face of potentially serious harm in the absence of scientific proof. The precautionary principle is not a matter for science alone, since it is a political and value laden statement expressing a fundamental shift in attitude of the general public to the environment . Taking a precautionary approach to the application of modern biotechnology to food and agriculture includes monitoring when necessary to indicate possible adverse effects, and the assessment of risk be modified in line with developing knowledge of the impacts of the technology and its applications.

The application of the precautionary principle to risk management of food biotechnology in practice, is found in the EC directive regulating the deliberate release of GMOs to the Environment. All releases of GMOs to the environment have to be given approval under the Deliberate Release Directive. The precautionary principle operates through a case-bycase, step-by-step assessment of the risks. Each GMO is considered individually and authorisation for placing on the market is only given once data has been collected from experimental trials which demonstrate that unrestricted use will have no adverse effects. Risk assessment needs to be based upon scientific calculations, but as these calculations are subject to uncertainties, the wider influences and social contexts affecting such calculations need to be acknowledged and incorporated into the process of risk analysis. The public's perceptions of risk are socially constructed, leading to a gap between the risk perceptions of the public and those of experts on a wide variety of hazards. While scientific estimates of risk fail to take account of social constructions of risk, the public's estimates are likewise based on factors other than scientific assessment.

The confidence in risk regulators, is also a factor in the public acceptance of risk, as are factors such as: dread, long term effects, and the concern that exposure to risk is involuntary. Where risk exposure is perceived as more voluntary, that is where the public perceive that they have the opportunity to make an informed choice, then such risk may become more acceptable. The public's assessment of these other factors, apart from scientific calculations, means that risk perceptions are socially embedded.At present, a number of these factors are not being addressed in the regulatory process. Research indicates that there is considerable lack of trust by consumers in both government and industry, yet it is these institutions which are responsible for introducing the techniques. Thus even if regulatory controls and risk analysis are properly conducted they might not be believed. The need is to link both the concepts of risk and trust. The incorporation of wider social concerns as articulated by different social actors should be included in risk analysis to produce a more socially embedded and acceptable process of risk analysis of the applications of modern biotechnology to food.

The key point of this modified scheme is that decisions are made incorporating both risk analysis and social impact analysis, building greater trust in and acceptance of the regulatory process. This will improve the quality of decision making and facilitate its expedition, which should at the minimum meet the same time frame as the previous scheme for risk analysis. In the revised scheme the regulatory and social streams will overlap and interact to produce a more widely negotiated set of criteria, which will frame the risk analysis. The decision is made by the risk manager, taking into account the social impact analysis and the lay perspectives. These perspectives will be continuously monitored and evaluated and will inform the decision making. The scheme will provide a more socially responsive, plural and accountable form of decision making.

Policy around the world

In 2000, countries that grew 99% of the global transgenic crops were the United States (68%), Argentina (23%), Canada (7%), and China (1%). Although growth is expected to plateau in industrialised countries, it is increasing in developing countries. There are two policy areas surrounding GM food:

- The standards and regulation of testing for food safety and;
- The requirements for labelling and tracibility of GM products in the food chain.

Countries vary on their approach to both these points. In the United States, genetically modified food is widely available. The left-wing environmental movement has attempted to challenge the adoption of gm crops by attacking the biotech industry and the limited governmental intervention in food processing and packaging. Efforts to force fear driven mandatory labeling and have been met with industry, consumer, and government indifference.

The Food and Drug Administration regulates the safety of all foods sold including the safety of GM foods. Under these regulations, substances added to food that do not meet the statutory definition of generally recognised as safe, and that are not pesticides, are classified as food or color additives and must be pre-approved before they may be marketed. Prior to marketing food additives, manufacturers are required to submit to FDA documentation demonstrating their safety and await approval for their use. For a list of the GM foods that have been regulated in this way see FDA's List of Completed Consultations on Bioengineered Foods.

Section 402(a)(1) of the Federal Food, Drug, and Cosmetic Act covers unintentional introduction of unsafe toxicants into foods. Producers of the foods are legally responsible to ensure that no substance occurs in the food that "may render" the food injurious to health. In order to comply with this part of the act GM companies subject new GM foods to safety testing in animal models.

Some people raise concerns that the safety of these novel food forms relies on the effectiveness of testing by food manufacturers. Defenders of the regulatory model point out that such a model has served the public well for many years with respect to non-GM food, drugs and food additives. They further point out that manufacturers data is subject to close scrutiny and that the law protects the consumers from manufactures of dangerous foods. It is thus in the manufacturers' best interests to ensure that the safety testing of GM foods is done adequately. Section 403(i) of Federal Food, Drug, and Cosmetic Act requires that a producer of a food product must reveal on a label "all facts that are material with respect to consequences which may result from use".

Under the act a consumer must be informed by labeling if a food derived from a new plant variety differes from its traditional counterpart with respect to a safety or usage issue. The FDA takes the view that the method of production of the new plant variety is not "material" to whether a food needs to be labeled under the act. The FDA's policy is that foods be considered on case by case basis and that labeling would be required on individual GM food products if these GM foods were materially different.

In Europe, a series of unrelated food crises during the 1990s outbreaks and foot and mouth disease) have created consumer apprehension about food safety in general, and eroded the public trust in government oversight of the food industry. This has further fueled widespread public concern about GMOs, in terms of environmental protection, health and safety of consumers and the right to make an informed choice. The apprehension might also be due to the novelty of GM foods, as well as cultural factors relating to food. The mishandling of the BSE crisis has left some consumers unwilling to consider "science" to be a guarantee of quality.

European consumers are demanding that their "right to know" the content and origin of the food they consume be respected. In a context of local food surplus where current GM food has little added nutritional value, many European consumers are wondering why any risk should be taken. However, as a result of the high quantity of GMO crops, the presence of GM in imported food products (shipments of grain for food, feed and processing for example), is now thought inevitable and largely unavoidable, and usually not mentioned.

For these reasons, the marketing of GM food is regulated in a manner that helps to provide the necessary levels of safety, transparency and reassurance. At the beginning of the 2000's, European officials insisted that new regulations were needed to "restore consumer confidence" in the technology. These new regulations required strict labelling and traceability of all food and animal feed containing more than 0.5% GM ingredients. Directives, such as directive 2001/18/EC, were designed to require authorisation for the placing on the market of GMO, in accordance with the precautionary principle.

One of the features of the European system is a comprehensive pre-market risk assessment, a system trying to provide means for products to be followed at each stage of their production and distribution, by both transmission of accurate information and labelling. This traceability is a means to implement post-market measures such as monitoring and withdrawals. This system is not only limited to GMO products but should encompass any food product ultimately.

The original EU rules for labeling of GM products were limited to products where transformed DNA and/or transformed protein are detectable, not to products that have been produced from GMOs but no longer appears to contain modified DNA and/or proteins. New rules for tracebility and labelling which came into force in 2004 also require labelling of highly refined products made from GM ingredients like oil and corn syrup, even though that the presence of recombinant DNA or protein cannot be proven. The labelling rules do not apply to products of microbial genetic engineering, so the cheese made with the help of GM-chymosin doesn't have to be labelled. Officials stress that while traceability facilitates the implementation of safety measures, where appropriate, it cannot and should not be considered as a safety measure.

In April 1998, a 5 year ban was pronounced on new genetically modified crops. At the end of 2002, European Union environment ministers agreed new controls on GMOs could eventually lead the 25-member bloc to reopen its markets to GM foods. European Union ministers agreed to new labelling controls for genetically modified goods which will have to carry a special harmless DNA sequence identifying the origin of the crops, making it easier for regulators to spot contaminated crops, feed, or food, and enabling products to be withdrawn from the food chain should problems arise. A series of additional sequences of DNA with encrypted information about the company or what was done to the product could also be added to provide more data.

Japan, like Europe, maintains labelling standards for GM food products. Japanese demand and assistance has led to a small effort to set up separate processing facility for non-GM soybeans in the U.S. Labelling is currently not required for GM food products sold in Canada. In 2005, a standing committee began work in the province of Prince Edward Island to assess a proposal to ban the production of GM foods within the province.

China is currently a producer of GM cotton. Research published in Science claims that Chinese farmers growing GM cotton use significantly less pesticides, reducing costs and improving farmer health. The Chinese government has also released safety certificates following field and laboratory testing allowing the cultivation of GM tomato, pimiento and a species of morning glory. Development of new GM crops for food is an active field of research in Chinese institutions.

In March 2002, China introduced biosafety rules that demanded strict labelling, extensive documentation and government approval for food shipments. Under these new rules, all soybean shipments from the United States were briefly interrupted until interim safety certificates could be acquired. In 2004 the Chinese Ministry of Agriculture announced its intention to assess the safety of GM rice lines developed by Chinese institutions for insect, disease and herbicide resistance. With government approval the crops may be planted as soon as spring 2006.

Agriculture officials from developing and other economically disadvantaged nations are receiving training courses on GMO at the American Agriculture Department, with instruction in the WTO rules on GM products and benefits of biotechnology. U.S. industry groups are also providing "technical assistance" to fund initiatives that promote "science-based and transparent biotechnology regulations" in countries such as China.

When, for instance, the U.S. Trade Repesentative launched a WTO case against the European Union over its moratorium on GM product approvals, he was flanked not by the heads of the U.S. corporations that stood to benefit from the hoped for improvement in market penetration, but by scientists who came, at least originally, from India and Kenya and by a 'small farmer' from South Africa .

There is good reason for scepticism about some of the pro-industry 'faces' being flown in from the global South. In some cases 'small farmers' supposedly leading a 'hand-to-mouth existence' in Asia or Africa have turned out not to be subsistence farmers at all. Some have been groomed by Monsanto and appear to be reading carefully scripted statements. Independent scientists, websites and lobby groups claiming to represent those in developing countries can also prove not to be as independent or representative as they claim.

7

Transgenic Technologies

For several decades, the ability to manipulate genes in organisms such as yeast, fruit flies, and nematodes has produced important insights into the regulation of a wide variety of complex biological processes. Limitations in the use of such organisms for research in neuropsychopharmacology arise from the marked organisational differences between the mammalian brain and the systems that govern behavior in these organisms. By contrast, a substantial degree of homology exists in the organisation of the central nervous system (CNS) and in the complement of genes expressed across mammalian species. Currently, the mouse genome is by far the most accessible mammalian genome to manipulation. Procedures exist in the mouse for introducing new genes, expressing elevated levels of endogenous genes, and eliminating or altering the function of identified target genes.

Mutant mouse models may be used for a number of purposes relevant to neuropsychopharmacologic research. For example, the impact of genetic mutations on the behavior of mutant mice may be examined, providing insights into the functional significance of particular gene products. In some cases, the manifestations (phenotypes) of these mutations may resemble features of human neuropsychiatric diseases, providing animal models for studying neural processes relevant to such disorders. Furthermore, as genes that confer susceptibility to human diseases are identified, it will be possible to introduce corresponding mutations into the mouse genome, generating useful models for studying disease pathophysiology and treatment. Finally, genetic models will be useful for investigating mechanisms through which nonselective drugs influence neural function and behavior. For example, the contribution of a particular receptor subtype to the actions of a nonselective drug may be examined by studying its actions in animals with targeted loss-of-function mutations of that receptor gene.

The development of transgenic and gene targeting technologies is significantly enhancing understanding of cellular and molecular functions of genes and their contributions to neural processes relevant to clinical disorders. In some instances, novel roles for receptor subtypes have been revealed by the observation of unexpected phenotypic abnormalities in mutant animals. Mutant strains are also providing models for studying the pathophysiology and treatment of particular neuropsychiatric diseases. In addition, some mutant mouse models are useful for investigating the mechanisms of action of psychoactive drugs. Although mutant mouse models represent powerful tools in neuropsychopharmacology, they do not replace older methods of investigation. These models may be best used as components of integrated multidisciplinary research efforts spanning multiple levels of analysis.

Transgenic Techniques

The ability to insert an exogenous (or foreign) gene into the mouse genome by direct injection into the pronuclei of zygotes was achieved just two decades ago. The term transgenic was applied to mice expressing exogenous DNA that had been produced using this technique. With this method, the gene of interest is inserted into a random locus in the mouse genome, and is expressed "in trans", i.e., not in its usual genetic locus. The techniques required for intro-ducing transgenes into the mouse genome have been highly refined, permitting their widespread use. Since the development of this technique, many thousands of lines transgenic mice have been generated, and it has been the most widely utilised technique of genetic manipulation mice.

Production of Transgenic Mice

Techniques for producing transgenic mice involve the microinjection of DNA constructs into fertilised mouse eggs. DNA constructs used for the generation of transgenic mice typically consist of a gene of interest located 3′ to promoter sequences selected to produce a desired distribution of gene expression. The maximum length of the DNA sequence that may be successfully incorporated into the mouse genome is not known, and up to 70 kilobase (kb) DNA fragments have been successfully integrated.

The transgene is linearised and purified from prokaryotic vector sequences. For optimal integration efficiency, about 1 to 2 picoliter (pL) of DNA at a concentration of 1 to 2 ng/μL (corresponding to a few hundred molecules of a 5-kb DNA fragment) is microinjected into the male pronucleus of a fertilized mouse egg. Although labor intensive, direct injection of DNA into the pronucleus results in much higher rates of integration of transgenes than other known methods of transformation. After microinjection, the embryos are surgically transferred into the oviduct of pseudopregnant mice. Pseudopregnant females are generated by matings with vasectomised males. The act of copulation initiates the endocrine changes of pregnancy, providing a suitable uterine environment for the survival and implantation of the transferred embyros. The foster mothers give birth 19 to 21 days after oviduct transfer.

For genotyping, DNA is typically isolated from mouse tail biopsies and screened for the presence of the transgene by Southern blotting or polymerase chain reaction (PCR). Typically, about 20% to 40% of the mice that develop to term possess the transgene. In the majority of cases, integration of the transgene occurs during the one-cell stage, so that the transgene is present in every cell of the transgenic mouse. Integration usually occurs at a single random chromosomal location, and, for reasons that are not fully understood, there are usually multiple copies of the transgene inserted as head-to-tail concatamers. Mice identified to possess the integrated transgene are referred to as founders. The founders are typically used in a breeding strategy to produce animals that are homozygous for the transgene insertion.

Uses of Transgenic Mice

Because transgenic mice often possess multiple copies of the transgene, this method can be used to produce animals with increased levels of expression of particular genes, i.e., mice that "overexpress" genes of interest. In addition, it can be used to express altered forms of a gene product in the distribution of the endogenous gene. One example is a transgenic line bearing a transgene composed of the Ca^{2+}/calmodulindependent protein kinase α subunit (CaMKIIα) promoter driving expression of a mutant form of CaMKIIα that conferred Ca^{2+}-independent activation. These mice exhibited an increased stimulation threshold for the induction of synaptic plasticity in the hippocampus, as well as deficits in spatial memory. Studies of these animals led to an enhanced understanding of the role of

CaMKII_ in synaptic plasticity and spatial memory acquisition.

In many cases, it is desirable to express a gene with an anatomic distribution that does not mirror its native expression pattern in the mouse. Such ectopic expression of a gene may be achieved using a transgenic construct in which the gene of interest is preceded by promoter elements that direct expression in an anatomic distribution characteristic of another gene. An example of this approach is a transgenic line in which the D1 dopamine receptor promoter was used to drive expression of a cholera toxin subunit (which constitutively activates Gs) in cells that express D1 dopamine receptors.

Studies of these animals revealed that chronic overstimulation (by constitutively activated Gs) of forebrain neurons expressing D1 receptors results in an abnormal behavioral phenotype that was likened to human compulsive behaviors. For most genes, the promoter elements necessary to reproduce the native patterns of expression are not well defined. A useful approach for identifying important promoter elements for genes of interest involves the generation of transgenic mice in which putative promoter sequences are used to direct expression of reporter genes, whose expression is readily determined in brain tissue. Comparisons may then be made between the pattern of reporter gene expression and that of the gene of interest.

It is also possible use transgenic approaches to reduce the expression of a particular gene product. This may be achieved using "dominant-negative" mutations, mutations that induce loss of function of a gene product when expressed in the heterozygous state. For example, transgenic constructs may be designed to express antisense RNAs that hybridize to native messenger RNA (mRNA) sequences, thus decreasing production of the gene product of interest. Alternatively, the function of gene products that aggregate into multimeric complexes may be disrupted by dominant-negative mutations that produce dysfunctional subunits of the complex. The most prevalent approach used to generate loss of function mutations, gene targeting, is described in the next section.

Interpretation of Transgenic Mouse Phenotypes

An important factor that frequently complicates the interpretation of studies with transgenic mice is the difficulty that may be encountered in achieving a desired anatomic distribution of transgene expression. Promoter elements are often quite large, and additional regulatory elements may at times be located great distances from the gene of interest. In addition, the site of integration often affects the pattern and level of transgene expression, so that founder mice generated with a common targeting construct may display different expression patterns. It may therefore be difficult to accurately duplicate a promoter's endogenous pattern of gene expression in the setting of a transgenic mouse.

Commonly, expression patterns are assessed in multiple founders, and those with the most appropriate transgene expression would then be selected for a particular experiment. Several additional considerations in the interpretation of phenotypes in transgenic mice warrant mention. For example, the number of copies of the transgene incorporated into the genome varies between founder mice. In some cases, concatamers can be unstable and susceptible to deletion of one or more copies of the transgene. In addition, the integration of the transgene may occasionally disrupt an endogenous gene. This could lead to the development of a phenotypic abnormality unrelated to the function of the transgene-this is estimated to occur in 5% to 10% of transgenic mice. This possibility may be assessed by determining whether similar phenotypes are present in animals derived independently from different founders because the likelihood of two founders possessing the same transgene integration site is minimal.

Gene Targeting Procedures

A mutational approach has proved to be invaluable to investigators examining the roles of gene products in complex biological processes within prokaryotic and cultured eukaryotic cells. Recently it has become possible to apply this approach to a mammalian system. Gene targeting procedures enable the precise (site-specific) introduction of a mutation to one of the estimated 100,000 murine genes. Typically, mutations have been designed to eliminate gene function, resulting in the generation of "knockout" or "null mutant" mice. The introduction of mutations that produce more subtle alterations in gene function has also been achieved. Two major developments have made gene targeting experiments feasible: (a) the generation of totipotent embryonic stem (ES) cells, and (b) the elucidation of techniques to achieve homologous recombination in mammalian cells.

ES cells are derived from 3.5-day-old mouse embryos, at the blastocyst stage of development. Blastocysts are cultured individually under conditions that permit the proliferation of the inner-cell mass cells, which are those cells that would normally become the fetus. These cells are then disaggregated, and individual ES cells clones are grown. Under optimal conditions, ES cells retain the ability to contribute to all of the tissues of the developing fetus. The derivation of ES cells was pioneered using embryos derived from the 129/Sv strain of mice, a strain that has been most commonly used in studies of early embryonic development. Although this mouse strain is not ideal for the study of behavior, most ES cell lines in current use are 129/Sv-derived.

Homologous recombination is the process by which a mutation is targeted to a precise location in the genome. A targeting construct is generated that typically consists of a target gene sequence into which a loss-of-function mutation has been engineered. Most targeting constructs are designed to achieve homologous recombination events in which recombination at the target locus results in replacement of native target sequences with construct sequences.

In mammalian cells, fragments of DNA preferentially integrate into the genome at random locations, at rates that greatly exceed homologous recombination. Therefore, targeting constructs are designed for use in selection strategies that enrich for ES clones in which homologous recombination has occurred. In the commonly used positive-negative selection strategy, a portion of a protein-coding exon is replaced by sequences that confer resistance to the drug neomycin. This mutation serves two purposes: (a) to inacti-vate the gene product, and (b) to provide a marker that enables the selection of cells that have integrated the construct. This exogenous DNA fragment is flanked by regions of DNA that are homologous to the native gene. Adjacent to one of these homologous regions is a gene encoding thymidine kinase. Treatment with the drug ganciclovir will kill cells that express this gene.

The targeting construct is typically introduced into ES cells by electroporation. In this step, cells are subjected to an electric current that facilitates the internalisation of the DNA construct. Those cells that failed to incorporate the targeting construct are killed by the addition of neomycin to the culture medium (positive selection). The majority of the remaining cells have incorporated the entire DNA construct (including the thymidine kinase gene) at random sites throughout the genome. By contrast, during homologous recombination, nonhomologous regions of the construct that are not flanked by homologous sequences are excluded from the integration event. Therefore, homologous recombinant clones will not contain the thymidine kinase gene.

Thus, the addition of a second drug, ganciclovir, will selectively kill cells that have randomly incorporated the construct (negative selection), thereby enriching for targeted clones. ES cell clones that survive this double drug selection are then screened for homologous recombination by PCR or Southern blot analysis. The homologous recombinant clones, which are heterozygous for the introduced mutation, are then used to generate chimeric mice. Following the isolation of homologous recombinant ES cell clones, these cells are microinjected into the fluid-filled blastocoele cavity of 3.5-day-old embryos at the blastocyst stage.

The injected embryos are then surgically transferred into the uterus of pseudopregnant females. These animals will then give birth to chimeric mice, which are derived partly from the injected ES cells and partly from the host embryo. For example, ES cells derived from a brown strain of mice are often injected into embryos derived from blackC57BL/6 mice, resulting in chimeras with coats containing blackand brown patches. The extent to which the ES cells have colonized the animal may be roughly approximated by the extent of the brown contribution to the coat. It is most important that ES cell derivatives colonize the germ cells of the chimera, so that the targeted mutation can be propagated to subsequent generations. If chimeras are mated with C57BL/6 mice, then the germ line transmission of ES cell-derived genetic material is indicated by the generation of brown offspring. Half of these brown mice will be heterozygous for the targeted mutation. These heterozygous mice are then bred to produce homozygous mutant mice that completely lackthe normal gene product.

Uses of Gene Targeted Mice

Studies of null mutant mice provide novel insights into the functional roles of neural genes and, in some cases, animal models relevant to human neuropsychiatric disorders. An illustrative example is a recent study of mice lacking the hypothalamic neuropeptide orexin. Observations of homozygous mutant mice revealed an unanticipated phenotypic abnormality. The mutants displayed frequent episodes of inactivity characterised by the sudden collapse of the head and buckling of the extremities. Subsequent electroencephalogram (EEG) analysis revealed these episodes to be similar to narcoleptic attacks observed in humans and in a strain of narcoleptic dogs. Moreover, a mutation of an orexin receptor gene was found to underlie the canine syndrome. Thus, the null mutant phenotype revealed a novel role for orexin in sleep regulation. In addition, this line of mice represents an important animal model for examining the pathophysiology and treatment of narcolepsy.

Another example illustrates the potential utility of null mutant mice to uncover mechanisms underlying the behavioral effects of psychoactive drugs. The nonselective serotonin (5-hydroxytryptamine, 5-HT) receptor agonist m-chlorophenylpiperazine (mCPP) interacts with several subtypes of 5-HT receptors. Although this compound typically reduces locomotor activity in rodents, it produced a paradoxical hyperlocomotor response in a line of 5-HT2C receptor null mutant mice. This response to mCPP was blocked by pretreatment with a 5-HT1B receptor antagonist, indicating that the absence of 5-HT2C receptors unmasked a hyperlocomotor effect of mCPP on 5-HT1B receptors in mutant mice. These results provide a model whereby genetic endowment may contribute to the development of a paradoxical drug response. When a compound alters the function of multiple gene products with opposing influences on behavior, then mutations or allelic variation of these genes may lead to paradoxical effects.

Although gene targeting techniques are most commonly used to generate animals with null mutations, subtle mutations may also be introduced that alter, but do not eliminate, function. The benefits of such an approach are highlighted by a mutation of a gene encoding the α_1 subunit of the γ-aminobutyric acid A (GABAA) receptor. The mutation produced a single amino acid change, rendering the α_1 subunit -containing sub-population of GABAA receptors insensitive to benzodiazepines, without affecting their responsiveness to GABA. The resulting animals exhibited reduced sensitivity to the sedative and amnestic effects of diazepam, but no change in sensitivity to the anxiolytic-like effects of this drug. These results indicate that benzodiazepine site ligands devoid of activity at α_1 subunit-containing GABAA receptors may act as anxiolytics devoid of some of the side effects typically associated with benzodiazepines, a prediction borne out by a recent report of the behavioral effects of such a compound. These insights would not have been obtained using a conventional gene targeting approach, because a null mutation of the α_1 subunit gene would profoundly perturb brain GABA signaling.

Interpretation of Targeted Mutant Phenotypes

In interpreting behavioral phenotypes, attention must be paid to the effects of genetic background. The phenotypic consequences of many targeted mutations may be influenced by modifying genes that differ among various inbred strains. In some cases, phenotypic abnormalities have been lost when mutations were bred to a new genetic background . It may therefore be useful to examine the persistence of mutant phenotypes in the context of several genetic backgrounds. In one example, three groups independently generated lines of mice with null mutations of the 5-HT1A receptor subtype. Interestingly, although each group placed this mutation on a different genetic background, all observed enhanced anxiogenic-like behaviors in the mutant lines. Thus, particularly strong evidence is provided for a contribution of the 5-HT1A receptor to the regulation of anxiety.

Another potential problem arises from the common use of ES cells derived from 129/Sv mice. Mice of this strain are susceptible to structural abnormalities of the CNS, such as agenesis of the corpus callosum, and are impaired in several behavioral assays. This potential problem may be addressed through breeding programs to place targeted mutations on different inbred backgrounds, and by the generation of ES cell lines derived from other inbred strains. It has been recommended that the C57BL/6 and DBA strains be used as standards, due to the extensive body of data relating to the behavioral characterisation of these strains.

In addition to the above strain considerations, the standard application of gene targeting technology has several inherent limitations. The null mutations engineered into knockout mice are typically constitutive, i.e., they are present throughout embryonic and postnatal development. Therefore, the potential for developmental perturbations is a major caveat to the interpretation of mutant phenotypes in adult animals. It may be difficult to determine whether a mutant phenotype reflects a normal adult role for the targeted gene or an indirect effect of the mutation attributable to perturbed development. Such an effect may lead to an overestimation of the functional significance of the gene product in the adult animal.

Conversely, if significant compensation for the loss of a gene product occurs during development, then the severity of the mutant phenotype may underestimate the functional significance of the gene product. The nature of such compensatory mechanisms and the extent to which they exist may be difficult to determine. The above considerations also pertain to the analysis of transgenic mice carrying constitutive mutations. Another limitation of the standard gene targeting technology is that the mutations are ubiquitous, present in all of the cells of the animal. Thus, if a neural gene of interest is also expressed in peripheral tissues, then the absence of the gene product peripherally could lead to a lethal or altered phenotype, independent of its neural role. Moreover, for genes that are widely expressed in the CNS, it may also be difficult to anatomically localise the brain region(s) that underlie the mutant phenotypes.

Conditional Mutations

New technologies are under development for circumventing the limitations of standard gene targeting approaches by creating mutations that may be induced in adult animals and/or restricted to particular brain regions. Although these strategies are not yet in widespread use, it is likely that rapid advances in this area will lead to an exponential increase in the generation of such "conditional mutations" over the next several years.

Cell Type-Specific Mutation Strategies

When a null mutation of a gene results in a mutant phenotype, limitations in the interpretation of that phenotype can arise because the gene is inactivated in every cell in which it was expressed in the mouse in the wild-type (WT) state. Therefore, the observed abnormal phenotype may arise from the absence of the functioning gene product in a peripheral organ system, the peripheral nervous system, or the CNS-i.e., in any of those regions in which the gene is normally expressed. It is possible that the absence of a gene product in the periphery may lead to embryonic lethality, precluding use of the mutant for the studies of neural function.

For genes that are widely expressed within the CNS, it may be difficult to identify neural circuits through which mutations produce behavioral perturbations. The ability to inactivate genes in restricted subpopulations of the cells that normally express them will be a valuable asset in studies to uncover the neural mechanisms underlying neural phenotypes. Recent efforts have focused on a mutational strategy developed to exert spatial control over the pattern of expression of genetic changes introduced into mice. This approach utilises somatic cell recombination rather than germ cell (or embryonic stem cell) recombination to inactivate a gene in restricted populations of cells or tissues. In this approach, a tissue-specific promoter is used to direct expression of one of the site-specific recombinases to limit gene inactivation to only those cells expressing the recombinase. The two recombinase systems that have been utilised for genetic manipulation in mice have been the Flp-frt system from yeast, and the Cre-lox system from bacteriophage P1, with the large majority of reports using this technique utilising the Cre-lox system.

Cre (cyclisation of recombination) recombinase is a 38-kd protein from bacteriophage P1, which recognises and catalyses reciprocal DNA recombination between two loxP (locus of crossing over of P1) sites. The loxP site is the 34-

base pair (bp) recognition sequence for Cre composed of a palindromic 13-bp sequence separated by a unique 8-bp core sequence. A gene or gene segment with flanking loxP ("floxed") sites will be excised by homologous recombination in the presence of Cre, leaving a single loxP site marking the point of excision and re-ligation of upstream and downstream DNA.

This approach, then, involves generating two independent lines of mice-a line bearing loxP sites, and a transgenic line in which Cre expression is driven by a tissue-specific promoter. Animals with a gene or gene region of interest flanked by loxP sites (floxed) are generated by gene targeting. Because the loxP sites are relatively small and placed in intronic regions, they do not typically interfere with normal gene transcription. Of course, WT patterns and levels of expression need to be documented in these floxed mouse lines, because inadvertent placement of lox sites into promoter elements or RNA splice sites could disrupt gene expression. The Cre mice are most commonly generated by creating a transgenic line of mice in which Cre expression is driven by a tissue-specific promoter.

Variability in transgenic expression patterns requires several lines of Cre mice that need to be generated and assayed for patterns of Cre expression. An alternative strategy is to use gene targeting procedures to place Cre under the control of an endogenous promoter. The advantage of this approach is that Cre expression should closely approximate theWTexpression pattern of the gene it is replacing because the original gene's promoter remains in its endogenous location. A potential disadvantage is that Cre may disrupt expression of the gene into which it has integrated. Once a line exhibiting the desired pattern of Cre expression is identified, it is crossed with an appropriate floxed line to commence a breeding strategy resulting in the generation of animals with a restricted pattern of gene inactivation.

Several lines of Cre mice have been reported in which expression is restricted to subpopulations of cells within the CNS. The first example of this approach was the inactivation of the glutamate receptor subunit NMDAR1 in CA1 pyramidal neurons of the hippocampus, with expression in other brain areas mostly intact. NMDAR1 is the predominant N-methyl-D-aspartate (NMDA) receptor subunit and is widely expressed in most CNS neurons. It had been previously demonstrated that widespread gene inactivation in NMDAR1 null mutants resulted in perinatal lethality. When the mutation was restricted to hippocampal CA1 neurons, animals were viable and exhibited impaired spatial learning and impaired plasticity at CA1 synapses. Thus, spatial restriction of neural mutations can be used to uncover particular brain regions or cell type through which gene inactivation alters behavior.

The utility of this approach for producing celltype – specific inactivation of genes is enhanced by the fact that the components of the system produced in one laboratory can be "mixed and matched" with components from another laboratory. That is, Cre lines generated for use with a particular floxed gene may also be used with other floxed genes when a similar pattern of gene inactivation is desired. Collaborative efforts to generate databases of Cre and floxed lines will speed and simplify the production of animals with restricted patterns of gene inactivation.

Inducible Mutation Strategies

The absence of a gene product throughout development complicates the interpretation of mutant phenotypes. Efforts are currently under way to overcome this limitation through the use of gene expression systems that may be induced in the adult animal. Strategies are under exploration for achieving this goal using a variety of compounds, such as tetracycline, steroid receptor antagonists, and ecdysone to induce gene expression. Although these approaches have yet to be optimised for general use, this development is likely to be close at hand. The tetracycline system has been the most utilised and best developed approach to inducible gene regulation.

Since the introduction of the Tet system by the Bujard laboratory in 1992, many laboratories have validated the utility of this approach to inducible gene regulation, and many refinements/improvements in the system have been introduced. This system is based on the regulatory elements of a tetracycline resistance operon of Escherichia coli, in which the transcription of tetracycline resistance genes is negatively regulated by the tetracycline repressor (tetR).

When tetracycline is present, tetR binds the tetracycline and loses its capacity to bind to the operator sequences (tetO) located within the promoter of the tetracycline resistance operon, and transcription is permitted. By creating a fusion protein composed of the tetR and a potent transcriptional activator, VP16, a tetracycline-dependent transactivating factor (tTA) was produced that retained the DNA binding and activation specificity of the tetR. The desired regulatable gene of interest is placed under tetO plus a minimal promoter (Pmin), that contains the basic promoter elements required for transcription in all cell types. Activation of this system requires the binding of the tTA to the tetO operator sequence.

The presence of tetracycline, or other suitable ligand such as doxycycline, prevents tTA from binding to tetO and activating transcription of the gene of interest. This is referred to as the tet-off system -that is, when tetracycline is present, transcription is off. A tet-on system has also been developed, in which tetracycline induces transcription of the gene of interest. It utilises a reverse tetracycline transcriptional activator (rtTA), designed so that it would bind to tetO and activate transcription only in the presence of tetracycline-related compounds. Doxycycline is most frequently used because it is a potent regulator in both the tet-off and tet-on systems, and can be easily supplied to mice through their water or food supply.

The tet-off and tet-on systems are binary systems-i.e., they require two genetic elements to be introduced into mice. First, a tissue-specific promoter can be used to express tTA or rtTA in a region or cell-type specific manner; then the gene of interest is inserted behind tetO and a minimal promoter. This can be achieved by creating two separate transgenic lines of mice and then cross-breeding to produce bigenic lines. In these lines, expression of the gene of interest may be induced by doxycycline (tet-on) or by the discontinuation of doxycycline treatment (tet-off). For example, the tet-off system has been used to investigate the effects of the transcription factor ΔFosB on psychostimulant responses. A line of mice was generated in which expression of a ΔFosB transgene was suppressed by continuous doxycycline treatment throughout development.

Discontinuation of treatment in adult animals led to overexpression of the transgene in the nucleus accumbens and to augmentation of the rewarding and locomotor stimulant properties of cocaine. The utility of the tet-on system has also been demonstrated. For example, a line of mice was developed to examine the role of the Ca^{2+}-activated protein phosphatase calcineurin in synaptic plasticity. Treatment of these animals with doxycycline induced calcineurin overexpression in restricted forebrain regions, associated with deficits of neuronal plasticity and spatial learning. Rather than generating regulatable gain of function mutants with the Tet system, regulatable loss of function mutants can also be generated by combining the Tet system with the Cre-lox system. In this arrangement, a celltype -specific promoter drives rtTA expression and Cre is linked to tetO and a minimal promoter. In the presence of doxycycline, Cre is expressed in the cell type specified by the promoter used to drive rtTA expression, and somatic cell recombination excises floxed DNA fragments in those cells-achieving an inducible cell-type-specific knockout.

In these inducible knockout mice it must be remembered that although the excision of the floxed gene can be induced relatively quickly, the appearance of any phenotype resulting from the absence of the gene product will occur gradually, depending on the degradation rate of the relevant mRNA and the half-life of the protein. Another important limitation of strategies utilising the Tet system relates to the inherent "leakiness" of the tetO operator; i.e., low levels of unwanted gene expression may occur during periods in which gene expression is expected to be turned off. This may be problematic when the inducible transgene is toxic or has significant effects even when expressed at very low levels. Recent findings with tetracycline controlled transcriptional silencers indicate that it may be possible to modify the tet system to substantially reduce unwanted gene expression.

Transgenic Crops

Conventional plant breeding has been extremely successful and increased financial support for plant breeding will be necessary if plant breeding is to both maintain and improve crop yields. However, there are some limitations inherent in conventional plant breeding such as lack of practical access to useful germplasm due to sexual incompatibility barriers or undesirable linkage block and concomitant time lags in incorporating useful genes into existing varieties. Indeed, many transgenic approaches to crop improvement arise from a lack of suitable conventional approaches to dealing with a particular agronomic problem or need. For many pests and pathogens which seriously limit agricultural productivity, transgenic approaches may provide new options where current options are lacking in their efficacy or existence.

Transgenic approaches can therefore be of use for a broad range of crops and areas where there are limited options available through conventional breeding e.g. nuclear male sterility, improved heterosis breeding, reducing toxic compounds, herbicide tolerance, generating novel resistance genes. Transgenic approaches have considerably broadened the range of genepools which are now accessible for crop improvement purposes. For instance, the application of useful gene transfer from microorganisms through genetic engineering techniques range from the introduction of vaccine antigen genes to aluminium tolerance genes to food plants. Isolated plant genes can now be usefully transferred between sexually incompatible crop plant species. In the context of ongoing debates regarding transgenic crops, public funding agencies should not forget that modern biotechnologies such as plant tissue culture, molecular genetic mapping and marker assisted selection could still have a major impact on any conventional crop improvement approaches which decide to limit themselves to the genetic variation accessible within the primary to secondary genepools.

Biosafety Assessment

Biosafety assessment requires that risks, benefits and needs be given a balanced assessment in relation to transgenic organisms. Many opponents of plant biotechnology cite biosafety as the key risk based issue for the more stringent regulation of transgenic plants. At one end of the extreme, environmental groups are now calling for a moratorium in some European countries on the planting of 'genetically modified foods'. The other end of the extreme would be no regulations regarding transgenic organisms. Much controversy and public scaremongering has been generated by antibiotechnology groups over the 'safety' of transgenic plants in relation to their perceived negative impacts on human health or the environment. A new development has been that many of the anti-biotechnology groups have conceded that there are benefits to be had from the application of genetic engineering for addressing human medical problems. However, it is the application of genetic engineering to agriculture that is now the key focus of attention of the anti-biotechnology interest groups. Indeed, civil society perceptions of agricultural biotechnologies in many countries are now distorted because of highly polarised lobbyist campaigns between the biotechnology -industry on the one hand and anti-biotechnology groups on the other.

The independent presence of public sector agricultural biotechnologists and scientists has often been lacking from this ongoing polarised debate. Similarly, many membership based organisations which are more broadly representative of civil society such as trade unions, producers organisations and farmers organisations have also not been involved in these debates. Switzerland is unique in having conducted a referendum to democratically assess public opinion regarding a number of genetic engineering issues. On June 7 1998 40% of the Swiss population voted in a referendum which called for a moratorium on the cultivation of genetically modified crops, bans on research on transgenic animals and on patenting of genetically modified organisms.

The socio-economic cost of non-access to some transgenic crops which may be of utility to farmers is rarely factored into risk-assessment procedures. Assessment of the immediate needs of different groups of farmers and consumers could feasibly become an integral component of biosafety riskassessment procedures, where costs and benefits could be seen in more social, rather than solely environmental terms exported from countries where food surpluses are a normality. The issue of who decides what level of risk farmers/consumers should be exposed to is also an important consideration for any countries development of biosafety regulations. This is an area where promoting the greater participation of organisations who actually represent the needs of different groups of farmers and/or consumers could be most appropriate. Many of the most vocal environmentalist or consumer organisations may not actually be very representative of farmers or consumers needs. Membership based NGOs may provide a better reflection of farmers and consumer views than single interest lobbyist NGOs.

The absence of a functional biosafety review system may negatively affects the local development and importation of new biotechnology products, and therefore farmers' access to potentially useful germplasm and technologies. On the other hand, a very stringent biosafety review system can also delay or prevent farmers' access. Indeed, the cost of a regulatory system for biosafety within any one country is an important factor which has implications for determining which farmers will ultimately have access to biotechnology products. High regulatory costs will also have an affect on what transgenic traits are ultimately to reach farmers and will further bias research towards wide rather than specific adaptation. High regulatory costs will select for only those traits which represent the greatest commercial gain to the developer of the variety. Such regulatory costs can be high - In the USA it can cost US$1 million to get a plant biotechnology product through the regulatory system. If such expensive regulatory systems are used in developing countries the cost will either bias all transgenic research towards meeting the needs of the wealthier sectors of society or plant biotechnology will remain primarily at the research stage.

Also, over-stringent biosafety regulations are likely to disproportionately benefit larger companies over smaller companies or public sector bodies by acting as significant 'barriers to entry' to certain markets. The higher the regulatory hurdles the higher the chance that competition will be stifled between companies and that any benefits of biotechnologies will not reach poorer farmers. For example, efforts by the US Environmental Protection Agency (EPA) to regulate some transgenic crop varieties as 'pesticidal plants' were opposed by 11 scientific societies representing 80,000 biologists and food professionals. Interestingly, an alliance between the Biotechnology Industry Organisation (BIO) and anti-biotechnology groups such as the Environmental Defense Fund were in favour of such types of regulation for differing reasons.

In determining whether low income farmers will have equitable access to certain types of transgenic crops, is of relevance therefore that FAO has recently indicated that any LMO that can be considered a pest of plants falls within the scope of the International Plant Protection Convention (IPPC) and will be subject to the provisions of the Convention. Indeed, the broader issue of whether future or anticipated socio-economic impacts of biotechnologies should be considered under the draft Biosafety protocol to the CBD is also currently a matter of debate. Many conventionally bred crops are by any biological definition transgenic as they contain genes or segments of chromosomes from totally different crop species. For instance most of the bread wheat currently under cultivation contains a large segment of a rye chromosome. Triticale is a conventionally bred transgenic crop containing full copies of both the rye and wheat genomes that was developed 60 years ago and is now grown on more than a million hectares in Canada, Mexico and eastern Europe.

Similarly most of not all crop varieties of sugarcane, tomato, potato, rice, maize, oat, sugarbeet, black currant, plum and many other highly bred crops contain genes or chromosome segments derived from different wild relative species. Wide crossing and embryo rescue technologies have been used by breeders for longer than genetic transformation as a means of transferring useful genes across plant species barriers. In a biological sense at least the inter species genetic modification of foods is not inherently new.

In the context of biotechnology risk assessment, there is a widely held scientific consensus that risk is primarily a function of the characteristics of a product (whether it is a purified chemical or a living organism to be field tested) and is not per se a function of the method of genetic modification. For instance, the US National Academy of Sciences concluded that assessment of the risks of introducing recombinant DNA-engineered organisms into the environment should be based on the nature of the organism and the environment into which the organism is introduced and not on the method by which it was produced. However, the current legal definitions of GMOs upon which most biosafety legislation is being constructed are largely 'process' rather than 'product' based in order to suggest that there is some fundamental distinction between the process of gene transfer resulting from sexual recombination and gene transfer resulting from genetic engineering. However, the same plant gene might feasibly be transferred by either conventional plant breeding or by genetic transformation and while the products of both processes for gene transfer could in theory both be genetically or phenotypically identical, one would be labelled a GMO and the other would not. For instance, there is little difference between mutagenesis-derived sulfonylurea or imidazolinone-resistant soybean, maize and oilseed rape and others crops with the same herbicide resistance genes transgenically introduced. Both transgenes and endogenous genes are, depending on their positions in the genome, likely to have similar rates or patterns of hybridisation.

Indeed, if precision gene swapping or knockout mutation approaches through homologous genetic recombination are perfected for plants in the same manner that they have been for mammals it is likely that such approaches will be applied to crop improvement. In instances where exactly the same genotype could be produced by either conventional mutagenesis or by genetic engineering, the 'process' based definition of a GMO will be increasingly difficult to sustain by any biological definition at least. Many of the biological phenomena which are often cited as unique biosafety issues for transgenic crops actually also occur in conventional plant breeding or other biological processes involving nontransgenic GMOs and in wild species. These include gene-silencing, paramutation, segregation distortion, evolution of neo-virulent pest/pathogen, evolution of fungicide or herbicide tolerance, gene-flow to wild relatives, allergenicity, etc. It is sometimes contended that transgenes may display novel 'emergent properties' when transferred to a novel genetic context.

However, conventional plant breeding and indeed agriculture itself also display emergent properties many of which have been beneficial to humanity. Standard plant breeding and selection procedures are equally applicable to the selection of the best transgenic lines from the range of lines generated through genetic transformation. The yield benefits derived from dwarfing genes which were a major factor in the Green Revolution are an example of beneficial emergent properties from conventional breeding that were selected for by plant breeders.

At a global level the agricultural area which is currently planted with transgenic crops as developed by genetic engineering techniques is small relative to the areas under conventional crop varieties and landraces. However it is increasing as some farmers adopt transgenic varieties. The total acreage of cultivated land in the world stands at over 1.4 billion hectares which is predominantly under conventional crop varieties. However for a few crops in a few countries (USA, China, Canada, Argentina) there are significant areas planted to transgenic varieties. In those countries where transgenic crops have been given regulatory approval, the proportion of crop area devoted to transgenic crops is increasing. In 1997 the global area under transgenic crops was 12.8 million hectares - a 4.5 fold increase from the 2.8 million hectares planted in 1996. In 1998 it is estimated that 30 million ha of transgenic crops were planted globally. The large increases in areas planted are currently limited to a few commercially important crops. For instance, transgenic soybean, maize, cotton and canola represented 85% of the global transgenic area in 1997, of which 75% was grown in North America.

The 'black box' approaches to conventional breeding based upon the generation of genetic variation followed by selection for useful phenotypes is in some ways more imprecise than single gene transfer through genetic engineering approaches. In crosses between weedy wild relatives and their crops, tens of thousands of genes are typically recombined in the progeny plants the exact phenotypic results of which are extremely difficult to predict. Radiation induced mutation breeding has been applied in plant breeding for decades with at least 20,000 know field tests of gamma radiation treated germplasm conducted in open field trials without any weedy or toxic mutants generated. The FAO/IAEA Division has been instrumental in technology transfer of radiation breeding approaches to developing countries. In spite of any uncertainties inherent in conventional or mutation breeding, such plant breeding has served society well without warranting the level of restrictions that are currently being perceived as necessary for all transgenic crops.

The technologies to routinely make transgenic plants have only been in existence for over a decade or so - which is of the same order of magnitude as the time required to conventionally breed a new plant variety. Much progress has been made regarding the technologies for generating transgenic crops through genetic transformation and these technologies are constantly being improved. Although some crop varieties and species may be more difficult to transform than others, few crop species are now considered to be ultimately untransformable and a significant number of cereal varieties can now be routinely transformed. In recognition of biosafety related concern over the use of antibiotic selection markers in the process for generating transgenic plants, improved transformation systems have been developed which allow the generation of marker free transgenic plants.

Rransgenic Foods and Health Issues

There has been much misinformation circulated in the popular media regarding perceived dangers to human health from use of transgenic crops or other GMOs. The recent public hysteria generated in the UK regarding transgenic foods stemmed from a non scientifically reviewed research report on feeding trials of rats with transgenic potatoes containing a lectin with known insecticidal properties. In this case, while the researchers report was subjected to 'political' peer review it has not yet been subjected to peer review in scientific journals and hence all of the conflicting conclusions to date regarding the raw data are considered premature. Many naturally occurring plant proteins and compounds can be anti-nutrients, toxic or allergenic. Indeed a significant number of crop species are toxic if not cooked or prepared properly to reduce or inactivate such compounds.

In most instances, standard procedures for assessing toxicity and allergenicity can equally be applied to conventional and transgenic varieties to identify those transgenics which are substantially equivalent to conventional varieties. Such standard testing procedures were sufficient to identify that a methionine-rich 2S albumin from the Brazil nut was allergenic to some people and hence was not as good a candidate as a non allergenic methionine rich sunflower seed albumin gene for gene transfer to improve the nutritional content of legumes. Selectable marker genes are used in constructing transgenic plants containing an associated transgene of interest, but are usually not required once the transgenic plants are produced. Existing biosafety regulations have been stringent enough to disallow corn-borer tolerant maize in Europe because of the extremely low risk of antibiotic tolerance spreading in bacteria in the rumen of cattle. Even though most selection markers in constructing transgenics are likely to pose little danger either to humans of the environment because of perceived consumer hysteria it is likely that future generations of final product transgenic plants will not contain selectable marker genes. This is because a number of quite efficient systems have now been developed for the development of 'marker free' transgenics. More acceptable marker systems which are not based on antibiotic or herbicide resistance genes are also being developed. Hence it is possible that biosafety considerations regarding such genes may gradually become less of an issue as improved 'marker free' transformation systems become available.

Consumers have a definite right to information and hence choice regarding which foods they purchase or eat. However, consumer information is based on the premise that the information provided to the consumer is of utility to the consumer in making an informed choice. For instance, knowledge of the biological or species composition of foods will be of use to those who suffer from allergies to particular foods or compounds. The USA requires labelling of transgenic foods that are substantially different from their unmodified counterparts, including foods that could contain a potentially allergenic compound such as a peanut protein or glutenins. Indeed, it is questionable whether the label 'genetically modified' conveys any information of utility to the typical consumer in terms of making an informed choice based on what is currently known about transgenic food. Nonetheless, such labelling is increasingly perceived as necessary by both the biotechnology industry and some governments.

An increasing number of the OECD countries are implementing provisions that require the labelling of genetically modified foods. For instance, the European Union has recently approved two directives which state that labelling must be applied to novel foods and their ingredients produced through the process of genetic engineering. Because of the commercial potential in diagnostic services for segregating transgenic and non-transgenic foodstuffs, major research efforts are now underway to develop such diagnostic methods. Indeed, the labelling of transgenic foods and products is likely to be welcomed by some companies as a means of unequivocally capturing added value. Corporate strategy will require that any of the biotechnologically generated value-added traits for food and feed, and even for industrial markets, will be stacked with a variety of input traits in a variety of combinations that will need to be segregated and identity-preserved to capture the enhanced value of the end products. This will lead to contract production and marketing systems for the resultant grains, oilseeds and their derivative products.

There are now differences between international trading blocs over requirements to label products developed using genetic engineering processes. While the USA does not require labelling of GMOs, the European Union and Japan, amongst others, have opted for labelling of food products or components produced by genetic engineering. It is likely that such requirements for GMO labelling will become a multilateral or bilateral trade negotiation issue over whether such labelling constitutes a trade barrier.

Indeed, because of consumer concern over transgenic foods in Europe there may be substitution effects generated in both directions regarding non-transgenic and transgenic foods. For instance, countries which do not have access to technologies to produce transgenic crops and whose exports are perceived to be threatened with substitution may find that the export of labelled non-GMO derived products may be a short term strategy to maintain markets while they devise a means of diversifying their exports in the face of competition from GMO derived products.

8

Use of Communication Technology in Biotechnology

Biotechnology is any technique that uses living organisms or substances from those organisms to make or modify a product, improve plants or animals, or develop microorganisms for specific uses. The key components of modern biotechnology are:

1. *Genomics:* the molecular characterisation of all species;
2. *Bioinfomatics:* the assembly of data from genomic analysis into accessible forms;
3. *Transformation:* the introduction of single genes conferring potentially useful traits into plant, livestock, fish, and tree species;
4. *Molecular breeding:* the identification and evaluation of desirable traits in breeding programs with the use of marker-assisted selection;
5. *Diagnostics:* the use of molecular characterisation to provide more accurate and quicker identification of pathogens;
6. *Vaccine technology:* use of modern immunology to develop recombinant DNA vaccines for improving control of lethal diseases.

Modern biotechnology will not solve all the problems of food insecurity and poverty. But it could provide a key component to a solution if given the chance, and if steered by a set of appropriate policies. These policies should guide:

1. increased public investments in R&D, including that in modern biotechnology;
2. regulatory arrangements that inform and protect the public from any risks arising from the release of genetically modified organisms (GMOs);
3. intellectual property management to encourage greater private-sector investment; and
4. regulation of the private seed and agricultural research sector to protect the interests of small farmers and poor consumers in developing countries.

Pro-poor policies can help expand agricultural R&D, including traditional and modern biotechnological research, in order to solve problems of particular importance to the poor. The problems of orphan commodities require particular attention. Given the high rates of return, more public support for agricultural R&D should be encouraged in most developing countries. Additional public financial support for R&D at the national, regional, and international levels would help to develop public goods the poor can afford.

Technology-transcending risks emanate from the political and social context in which the technology is used. They include concerns that biotechnology may increase the prosperity gap between the rich and the poor, and may contribute to a loss of biodiversity. Ethical concerns about patenting living organisms and moving genes between species also fall into this category. The principles and practices for assessing and managing technology-inherent risks are well established in several countries. They take into account the nature of the crganism, the familiarity of the product, any distinguishing features of the process by which a product was produced, and the environment into which it will be introduced.

A science-based assessment of these factors on a case by case basis, and identification of any concerns expressed by stakeholders, enable regulators to find out what risks may be associated with a particular product and to make appropriate recommendations. A regulatory system that enjoys the confidence of the public and the business and farming communities is essential for the effective use of biotechnology.

The purpose of intellectual property management is to protect local inventions and enable access to technologies developed elsewhere. Trade-related intellectual property rights (TRIPs) are a matter of ongoing concern within the World Trade Organisation. The present patent system favors those countries that have a strong innovation base. Despite much effort, no satisfactory system exists to recompense traditional owners and improvers of germplasm. The lack of intellectual property protection also constrains private-sector investment in developing countries.

The participation of the private sector is critical to the development and delivery of new biotechnology products. The enabling environment to encourage privatesector participation includes a regulatory system that accurately informs the public of the benefits and risks involved in the use of new technologies, a legal framework for protecting intellectual property; adequate infrastructure for power, transport, and telecommunications; a fair tax system and investment incentives; a skilled workforce, including a well-supported university sector; public funding for R&D; and incentives to establish innovative public-private collaboration and joint ventures at the national and international levels.

The successful application of modern biotechnology to the problems that cause undernourishment and poverty could be called a biosolution. The delivery of new biosolutions to the problems of food security and poverty will require continual policy development and actions at the national, regional, and international levels. These efforts will involve the following five areas:

1. determining priorities and assessing relative risks and benefits in consultation with the poor, who are often overlooked while others decide what is best for them;
2. setting policies that benefit the poor and that minimise technology-transcending risks that adversely affect the poor;
3. establishing an environment that facilitates the safe use of biotechnology through investment, regulation, intellectual property protection, and good governance;
4. actively linking biotechnology and information technology so that new scientific discoveries worldwide can be assessed and applied to the problems of food insecurity and poverty in a timely manner; and
5. determining what investments governments and the international development community will have to make in human and financial resources in order to ensure that biosolutions to the problems of food security reach the poor.

Biotechnology Policies

Modern biotechnology can enhance agricultural productivity in developing countries in a way that further reduces poverty, improves food security and nutrition, and promotes sustainable use of natural resources. But such benefits from biotechnology require policy action on a number of fronts. The small farmer in developing countries faces a variety of problems and constraints. Crop losses due to insects, diseases, weeds, and drought threaten income and food availability. Acid soils, low soil fertility and lack of access to reasonably priced plant nutrients, and other biotic and abiotic factors also contribute to low yields. Poor infrastructure and dysfunctional markets for inputs and outputs, along with lack of access to credit and technical assistance, add to the problems plaguing the small farmer. Solutions to these problems will benefit both farmers and consumers.

Although modern biotechnology cannot solve all these problems, it can provide a critical component to the solution if it is guided by appropriate policies. Four sets of policies are particularly important. Policies must expand and guide research and technology development to solve the problems of particular importance to the poor. These problems include diets with inadequate levels of energy, protein, and micronutrients, and crop losses due to biotic and abiotic factors. Research should focus on the crops of particular importance to small farmers and poor consumers in developing countries. Bananas, cassava, yams, sweet potatoes, rice, maize, wheat, and millet, along with livestock products, feature most prominently in the diets and production activities of the poor. Except for limited work on rice, bananas, and cassava, little biotechnology research currently focuses on helping the small farmer and poor consumer solve their productivity and nutrition problems. The prediction so often heard that the poor in developing countries are unlikely to benefit from modern agricultural biotechnology in the foreseeable future could well come true-not because the technology has little to offer but because it will not be given a chance.

There are three ways to expand biotechnology research for the benefit of the poor. First, allocate additional public resources to agricultural research, including biotechnology research, that promises large social benefits. Existing national and international agricultural research systems have to be strengthened or new ones built. Low-income developing countries currently invest less than 0.5 percent of the value of agricultural production in agricultural research, compared to about 2 percent in developed countries. Underinvestment is widespread despite high annual economic rates of return from investments in agricultural research.

A recent assessment of more than 1,000 research projects and programs found an average annual rate of return of 88 percent. Investments by the private sector are limited to research that permits a large enough profit from the returns. Nonetheless, privately funded research can still generate large benefits to farmers and consumers, as illustrated by a recent study of the distribution of benefits from the use of genetically modified (GM) soybeans in the United States. The private patent holders and private seed companies captured one-third of the total economic benefits, farmers and consumers gained twothirds. While private-sector agricultural research has increased rapidly in the industrialised countries during the last 10 to 15 years, it currently accounts for a small share of agricultural research in most developing countries.

Second, expand private-sector research for the poor by converting some of the social benefits of research to private benefits for the private sector. The public sector can entice the private sector to develop technologies for the poor by offering up front to buy the exclusive rights to newly developed technology and make it available either for free or for a nominal charge to small farmers. The amount of the offer could be determined on the basis of expected social benefits, using an annual rate of return normally expected from agricultural research, for example, 60-80 percent.

The risk of failing to develop the specified technology would rest with the research agency, just as it does when technology is developed for the market. The public sector offer would come due to the research agency that first develops the technology, but only when the technology is developed, tested, and made available. Both private-and public-sector agencies could participate in this research. Opportunities for collaboration between multinational life science companies and public-sector agricultural research agencies in both developing and developed countries might increase the probability of success. With necessary refinements, the arrangement proposed here should be of interest to international development assistance agencies.

The third way to expand biotechnology research to help the poor is to protect the intellectual property rights of a private research agency that develops a particular technology, for example, seed with infertile offspring, or that contracts directly with the farmer, in both cases forcing the farmer to buy new seed every season. This would make it easier for the private sector to recuperate the incomes needed to justify the research. But seeds with infertile offspring may be inappropriate for small farmers in developing countries because they pose large risks to food security. Existing infrastructure and production processes may not be able to keep fertile and infertile seeds apart. Small farmers could face severe consequences if they planted infertile seeds by mistake. Monitoring and enforcing contracts that prohibit large numbers of small farmers from using the crops they produce as seed would be expensive and difficult to do.

GM foods are not intrinsically good or bad for human health. Their health effect depends on their specific content. GM foods with a higher content of digestible iron are likely to benefit consumers with iron deficiencies. But the transfer of genes from one species to another may also transfer characteristics that cause allergic reactions. Thus, GM foods need to be tested for allergy transfers before they are commercialised. It was precisely such testing that avoided the commercialisation of maize with a Brazil nut gene.

GM foods with possible allergy risks should be fully labeled. Labeling may also be needed to identify content for cultural and religious reasons or simply because consumers want to know. Finally, labeling may be required to identify the production process itself when that, rather than any specific health risk, interests consumers. Failure to remove antibiotic-resistant marker genes used in research before a GM food is commercialised presents a potential although unproven health risk. Recent legislation in the European Union requires that such marker genes be removed before a GM food is deemed safe for consumers. Risks and opportunities associated with GM foods should be integrated into the general food safety regulations of a country.

Effective national biosafety regulations should be in place before modern biotechnology is introduced into a country's agriculture. Such regulations should be country-specific and reflect relevant risk factors. The ecological risks policymakers need to assess include the spread of traits such as herbicide resistance from genetically modified plants to plants that are not modified, and the buildup of resistance in insect populations. Seeds that produce infertile offspring may be an effective solution to the risk associated with cross pollination but, as mentioned earlier, they may be inappropriate for small farmers. The approach used to develop terminator seeds, however, offers great promise for the development of a seed that will avoid the spread of new traits through cross-pollination. The seed would contain the desired traits, such as pest resistance or drought tolerance, but each trait would be activated only after treatment with a particular chemical. Without treatment, the seed would maintain its normal characteristics. Thus, if a farmer planted an improved seed, the offspring would not be sterile; rather they would revert back to being normal seeds.

The farmer would then have the choice of planting the normal seed or bringing back the improved traits by applying a particular chemical. Contrary to the terminator gene, this approach complies with the principle of doing no harm. Both food safety and biosafety regulations should reflect international agreements and a society's acceptable risk levels, including the risks associated with not using modern biotechnology to achieve desired goals. The poor should be included directly in the debate and decisionmaking about their desire for technological change, the risks of that change, and the consequences of no or alternative kinds of change.

Recent mergers and acquisitions have resulted in increasing concentration among companies engaged in biotechnology research. The outcome of this growing concentration may be reduced competition, monopoly or oligopoly profits, exploitation of small farmers and consumers, and successful efforts to gain special favors from governments. Effective antitrust legislation and institutions to enforce the legislation are needed, particularly in small developing countries where one or only a few seed distribution companies operate. Effective legislation is also required to enforce intellectual property rights, including those of farmers to germplasm, along the lines agreed to within the frameworks of the World Trade Organisation and the Convention on Biological Diversity.

Modern biotechnology research may help reduce poverty, improve food security and nutrition, and make the use of natural resources more sustainable, only if it focuses on the problems and opportunities poor people in developing countries face and only if appropriate policies accompany it. Modern biotechnology is not a silver bullet, but it may be a powerful tool in the fight against poverty and should be made available to poor farmers and consumers.

Governing the GM Crop Revolution

Will developing countries adopt policies that promote the planting of genetically modified (GM) crops, or will they select policies that slow the spread of the GM crop revolution? The evidence so far is mixed. In some prominent countries such as China, policies are in place that encourage the independent development and planting of GM crops. Yet in a number of other equally prominent countries the planting of GM crops is not yet officially approved. The inclination of developing countries to promote or block the spread of GM crops can be judged by the policy choices they make in five separate areas: intellectual property rights (IPR) policy, biosafety policy, trade policy, food safety policy, and public research investments.

Policy Options toward GM Crops

If developing countries want to bring GM crop technologies into their farming systems, they may have to recognise some of the intellectual property rights claims of the private companies that have been developing GM crops. At one extreme, they might even adopt the U.S. approach and provide full patent protection. A somewhat less promotional policy could offer only plant breeders' rights as IPR guarantees, which entitle breeders to use protected varieties as an initial source of variation for the creation of new varieties as in the 1991 agreement of the

International Union for the Protection of New Varieties of Plants (UPOV). A still weaker approach would be to embrace an earlier 1978 version of UPOV, which preserves the privilege of farmers to replant seeds from protected varieties on their own farms. Weaker still would be to provide no IPRs at all for plant breeders.

In the area of biological safety, the most promotional policy toward GM crops would be to approve the use of these crops without any careful case-by-case screening for unwanted geneflow or damage to nontarget species. A less promotional approach would be to screen GM crops case by case but only for risks that can be scientifically demonstrated. A more cautious approach would be to hold crops off the market case by case even without proof of risk so long as some scientific uncertainties remained. The most cautious approach would be to assume risk in all cases because of the novelty of the GM process.

Consumer acceptance of GM crops in major importing countries continues to evolve. Assuming adequate consumer acceptance, a promotional trade policy toward GM crops would be to seek the import of GM plant materials and seeds without restriction and promote the planting of GM crops in hopes of cutting farm production costs and becoming a more competitive exporter. A more neutral approach would be to neither promote nor prevent the planting of GM crops and to treat GM seed and commodity imports the same way as non-GM imports. A more cautious trade policy approach would be to develop and implement a separate and more restrictive method for regulating and labeling the import of GM seeds or commodities compared with non-GM. A preventive trade policy choice would be to ban GM imports and block the planting of GM crops. If consumer acceptance of GM crops in international markets continues to weaken, such a ban on planting GM crops could be defended on trade grounds as a way to seek price premiums on the world market as a "GM-free" exporter.

GM crops currently on the market pose no new hazards to human health and to impose no additional inspection or labeling burdens on them. A less promotional approach would be to require labeling of some GM foods in the interest of a consumer's right to know but to make the labeling standards lenient enough so that a complete segregation of GM from non-GM commodities is not required. A still more cautious approach would be to impose mandatory comprehensive labeling for all GM foods in a manner that would require market segregation. A fully preventive approach would be to ban all GM foods or to label them in ways intended to stigmatise and prevent their use.

Developing countries must also make a range of agricultural research investment choices toward GM crops. At one extreme they might spend treasury resources to develop their own GM crops. As a second option they could invest only in the more limited goal of backcrossing GM traits developed by others into their own domestic germplasm. As a still more limited option they could allow their scientists to pursue backcrossing of transgenes into local varieties only if donors were willing to pay for it. At a preventive extreme they could decide not to spend any money, even donor money, on GM crop research.

Policy Choices

This system can be used to classify the actual policy choices toward GM crops that were made by governments in Brazil, China, India, and Kenya in 1999-2000. Whereas China opted for relatively permissive policies toward GM crops, Brazil, India, and Kenya have in most respects been more precautionary. In Brazil, India, and Kenya biosafety approval has emerged as the principal point of resistance against moving the GM crop revolution forward. This is a surprising discovery given the fact that biosafety approvals for GM crops have not been such a strong sticking point in the industrial world, given the traditionally weak agricultural biosafety policies of most developing countries, and given the potential biosafety benefits of some GM crop applications. It is particularly surprising that Brazil and India have moved so slowly on biosafety approvals for GM crops, given the significant state investments that are simultaneously being made in both countries to develop GM crops.

International pressures of four kinds help explain this pattern of caution in the developing world:

(1) environmental groups based in Europe and North America have used media campaigns, lawsuits, and direct actions to project into the developing world a tone of extreme caution toward GM crops;

(2) consumer doubts in Europe and Japan regarding GM crops have discouraged planting of those crops by developing-country exporters;

(3) the precautionary tone of the 2000 Biosafety Protocol governing transboundary movements of GM crops is reinforcing biosafety caution in the developing world; and

(4) donor assistance to developing countries in the area of agribiotechnology has often focused more on the possible biosafety risks of the new technology than on its possible agronomic or economic advantages.

A further spread of GM crops into the developing world will depend on more than just the availability of suitable technologies. It will also depend upon the future willingness of biosafety authorities in developing countries to give farmers permission to plant GM crops. This willingness, in turn, will likely depend as much on the external pressures and influences faced by these regulators as upon actual documented threats to biosafety from GM crops.

Digital Technologies and Digital Divide

Without doubt the Internet and other information and communication technologies (ICTs), such as faxes, video, digital radio, mobile phones, and satellite technologies, have helped people gain access to, process, respond to, and distribute information in a faster and more far-reaching way than ever before. This ongoing shift toward a more information-intensive society gives both added weight and new meaning to the old maxim, "information is power." ICTs at a minimum can enhance the livelihoods of the poor and improve market efficiency. As development tools in their own right, ICTs can also help lead to higher literacy rates through distance learning; gender equality through the empowerment of women who gain greater access to economic opportunities and civil society; sustainable development through easier dissemination of appropriate information; more balanced social relations through the greater accountability imposed on the powerful by the marginalised; and other global goods.

The importance of ICTs is precisely why concern has escalated over the "digital divide"-the gap in information and communication technologies that exists between technologically advanced and developing countries, poor and rich, men and women, and urban and rural areas. Africa claimed only 2.8 million Internet subscribers, Asia/ Pacific 75.5 mil-lion, Middle East 1.9 million, and South America 13.2 million. In 1998, according to the International Telecommunications Union (ITU), low-income countries had 6.2 personal computers and 45 fixed-line and mobile phones per 1,000 people, whereas high-income countries had 311 and 832, respectively. The digital divide is not only enormous-dwarfing even the per capita income gap ratio between high- and low-income countries-but it is increasing, according to a recent World Bank study. The authors of the study, Francisco Rodrigues and Ernest Wilson, III, estimate that even East Asia, which is adopting ICTs at a rapid pace, will take another 40 years to catch up with the developed world. Inevitably, access varies widely among and within developing countries as well. Only 11 of 53 African countries for which there are data have 10,000 or more Internet users; 17 have fewer than 1,000. Gender, income, age, and other disparities also abound. In Ethiopia, for example, 98 percent of Internet users in 1998 were university graduates and 86 percent were male.

As the Internet becomes the norm for commerce and information exchange in the developed world, the poorer countries will have little choice but to try to bridge the digital divide in all its forms. They will go online to become competitive in the new economy or forgo significant growth. The stakes involve not only social development and productivity growth, but also the burgeoning e-commerce sector, which is expected to reach between US$2 and 3 trillion in transactions in the next three years.

Rural Poor and Access to Information

The last few years have witnessed a flurry of activity as the development community has put forward position papers and proposals and launched pilot projects to include the poor in the information revolution. Projects have ranged from providing price information to farmers to boosting microcredit-based sales via the web. The hallmarks of this activity have been diversity and innovation in organisation, technology, and funding. Results are trickling in, and no formula has emerged, though some ingredients appear to be critical. According to Roger Noll, professor of economics and director of the Public Policy Program at Stanford University, "Technological change has greatly affected the optimal strategy for starting ICT service. In many cases, fixed wireless is better for new networks because of its lower cost and quicker availability. And because economies of scale are not very important in wireless, competition is often feasible right from the start. Another common ingredient is that western-style universal service is not a practical reality in developing countries, where much of the population is too poor to afford individual access. The focus instead should be on convenient access through sharing, such as with telecenters and pay phones."

This need to minimise cost but maximise rural access drives most ICT projects in developing countries. Not surprisingly, the Internet still often plays a lesser role than other technologies in the distribution of information in rural areas. Most farmers, for example, are years away from getting price information through the Internet because of cost and lack of human resources and adequate phone infrastructure in rural areas. But because a farmer's decisions about where, when, and how much to sell depend on the signals sent by national and international market prices, ICT proponents see the delivery of timely price information as one of the key benefits farmers can receive from the new information technologies.

Consequently a number of ICT projects focus on making it easier for agricultural producers to access prices. "Price information helps connect farmers to markets," says Mike Weber, professor and codirector of a food security program run by the Department of Agricultural Economics at Michigan State University (MSU), "and the more up-to-date the price information, the better for the farmer and the market." Weber's own group uses both low- and high-tech means to distribute prices. For the past ten years, the food security project has delivered weekly agricultural prices by newspaper and fax to government offices, the private sector, and nongovernmental organisations (NGOs) in Mozambique. MediaFax, a fax-based newspaper, provides the same information each week to a wider audience. NGOs make copies of the weekly price lists and distribute the information to farmers.

Some projects are devised to pay for themselves while providing developingcountry poor with access to ICTs for socioeconomic development. Such is the case with the projects facilitated by the International Institute for Communication and Development (IICD), which was established by the Dutch Ministry for Development Cooperation in 1997. In Jamaica, IICD and its partners are helping to put in place a comprehensive agribusiness information system that will enable small farmers, who are losing ground in the global marketplace, to get both print- and web-based information about markets, products, registration requirements, and technical assistance. This demand-driven system makes information available through a continually updated database on domestic agricultural production and marketing and through the Internet. Extension agents collect the information from producers and traders using palm-top computers. The information is then downloaded to PCs located in primary extension offices. These PCs are linked to the central database at the headquarters of the Rural Agricultural Development Authority.

The system is currently being tested with 80 small farmers and traders in two regions in Jamaica. The method for delivering information is still evolving. Currently large farmers and traders with access to the Internet get the information easily. Smaller farmers and traders generally visit the extension offices, where they can obtain standardised and customised reports in hard copy. If literacy is a problem, extension officers help interpret documents. A small subscription fee makes the project sustainable, and small farmers can pool their resources and subscribe as a group. The belief that ICTs have a role to play in the lives of small farmers has led to their introduction in many rural areas. But most projects remain relatively smallscale and not widely replicated because of cost, the multitude of approaches and local needs, and lack of interest from the private sector. These factors, and the fact that ICT introduction is still at an early stage in developing countries, make the impact of projects hard to judge.

Some serious social science research is needed to understand the institutional context that can make these technologies really useful and sustainable in specific rural settings and it is not the Internet that is always most important. Older technologies too, such as community radio, are opening up societies and allowing people to talk to each other. The telephone, of course, remains vital to people's well-being. Perhaps the best-known example of bringing telephones to the poor is GrameenPhone, the Grameen Bank's rapidly growing cellular phone business in Bangladesh. A little more than three years old, the venture has already put mobile phones in the hands of women in more than 1,200 Bangladeshi villages. At the same time, GrameenPhone has secured more than 50 percent of the national, primarily urban, mobile phone market in Bangladesh, thus helping to assure both its financial ability to serve rural areas and its technical ability to create a reliable urban network with which to link the rural population.

In the villages, GrameenPhone works on the same principle as the Grameen Bank's microloan program, giving rural women from landless households access to credit. A woman who has already established good credit with the Bank, whose house is located in a fairly central part of the village, and whose family has one member familiar with the English letters and numbers on a phone, can borrow the roughly $350 needed to purchase a solar-powered mobile phone. After a day's training, the woman is set to provide phone service to other villagers for a price. This access to technology not only generates substantial income for the "telephone woman," who on average earns $450 a year after expenses, but also provides villagers with access to information and services that would otherwise remain far outside their reach. The villagers, can contact medical help immediately; get prevailing market prices for the crops they grow, thus avoiding underpayment by opportunistic traders; engage in commercial activities that require quick or frequent access to timetables, regulations, or other market-related information; and easily keep in touch with family members living in the cities or abroad.

The arrival of the cell phone has also, for once, turned the social pecking order on its head as the relatively wealthy rely on the poor to keep in touch with the outside world. "Money, the old saying goes, speaks loudest, but it is user-friendly technology that begets money in the first place," says Nuimuddin Chowdhury, an IFPRI consultant on ICTs. "The GrameenPhone project has shown once again that the rural poor are among the most eager to innovate, and they could significantly improve their income if given access to ICTs, which are, moreover, environmentally clean technologies."

For developing countries, bridging the digital divide generally means bridging the gap between super-sophisticated technology on the one hand and local logistics and financial ability to pay on the other. Because satellites can bring ICTs to rural areas at a relatively low cost, some NGOs promote them to help fill the information and communication needs in developing countries.

Volunteers in Technical Assistance (VITA), a U.S.-based NGO, for example, links its satellite with ground terminals that enable users to send and receive stored emails four to six times a day as the satellite passes overhead. Kept in a community center or some other central location, each terminal is estimated to serve an area of roughly 10,000 people via the email access it gives to schools, clinics, NGOs, business people, and individual users. VITA's planned services include continually updated web indexes designed to meet the information needs of remote communities and a research service to maximise the impact of the 50-page limit that can be downloaded each day by each terminal. Costs for these services and for supporting the system will be defrayed by a relatively small annual fee per terminal and by proposed private-sector use of a portion of VITA's satellite bandwidth in exchange for technological production and support. Currently at the experimental stage, the project serves 25 terminals.

VITA plans to deploy 2,500 terminals throughout the developing world. "Basically we're trying to push the envelope to give the poor access to a culture of information," says George Scharffenberger, president of VITA. "Instant access to the Internet is exciting, but for a third to a half of humanity affordable access is still several years away. In the meantime, satellites can provide low-cost communication and information services to poor and remote areas while building critical information skills among teachers, health workers, people in small businesses, women, and youth." VITA's other partner in this venture, SATELLIFE, also runs a health information system, HealthNet, which works through a satellite-terminal system similar to VITA's and through modem-to-modem telephone links.

One of Africa's first sources for email, HealthNet also provides electronic publications, access to the web, and discussion groups, including one on AIDS, to about 10,000 health professionals worldwide. Locally owned and operated, HealthNet has been helpful in dramatic cases, such as containment of an Ebola virus outbreak in Gabon, as well as in more routine health care. It has succeeded in those countries where an adequate institutional structure exists or can be built to support ICTs, where training of technical staff matches investments in technology, and where a successful business model has been implemented to offset operating expenses.

As ICTs begin to play a greater role in development, concerns are growing that women will be left out of the picture. Available data show that women account for 25 percent of ICT users in Brazil, 17 percent in South Africa, 7 percent in China, and 4 percent in Arab states. All the gender faultlines are present in ICT access, with poor, uneducated, and older women particularly affected. To address these concerns, the United Nations Development Fund for Women (UNIFEM), ITU, and UNDP recently signed an agreement guaranteeing the inclusion of gender issues in their policy dialogue and decisionmaking about ICTs.

The success of some local-level projects says little yet about what should be the method and timetable for delivering universal access. Even the dilemma about whether to focus first on building a national infrastructure for instant access or on building ICTs around specific development problems remains unresolved. Some developing countries, such as Malaysia, have made commitments to rapid and widespread ICT introduction.

Others may move forward, at least in the near future, on two tracks: partnering with the private sector to install ICTs in urban areas and looking to donors, lowest-cost methods, and innovative schemes for subsidizing subscription fees to connect rural areas, with the two tracks sometimes overlapping.

9

Sustainable Use of Biotechnology

Urbanisation and economic development, and the rise in income and expectations associated with them, make increasing demands on the food production system. People want to eat a more varied diet, and they want and can afford to eat more animal products. Global meat production has increased dramatically in the last 50 years, quadrupling since 1950. Until about 1950 in the industrialised countries, and even today in many parts of the developing world, farmers who practiced a sustainable mode of farming integrated livestock rearing with food crop production. They rotated food crops with feed crops. People ate the former; farm animals ate the latter. Farmers spread animal wasteover the soil as fertilizer, and leguminous feed crops added nitrogen to the soil.

However, this integrated system is hard to sustain when demand for meat is high. Specialised animal production facilities, often located close to the population centers, feed animals soybeans and grains such as maize and sorghum. This system requires large areas of land just to raise food for animals. To understand why eating animal products puts additional demands on the food production system, also consider how matter and energy transfer from one organism to another in the ecosystem. In the food chain, plants are called primary producers.

When an animal eats the plants, only about 10% of the plant matter and energy ends up as part of the growing animal's body. The other 90% is "lost": The animal uses a lot as energy to fuel its digestion and to refashion plant material into animal material. So if humans eat plants directly, the transfer efficiency is about 10%. Now consider what happens if an extra step occurs between plant and human: Let's say a cow eats grain, and then a human eats the cow. The efficiency from plant to human via cow is 10% ÷ 10%, or 1%. So people must grow 10 times the amount of grain to supply a meat-eating human with the same energy and matter as would be needed if the person ate the grain directly.

In some cases modern methods have improved the efficiency of transfer from grain to animal. Efficiency is higher in feedlots and chicken farms, where animals are confined and fed optimal diets. Some people have argued that eating animal products wastes precious agricultural resources, but there is at present not enough economic demand for food grains to expand their production.

Despite ecological inefficiencies, meat production is rising rapidly. The existence of several successful farming systems based on animal husbandry confirms that under certain conditions the inefficient conversion of energy and protein is also a positive aspect of the food production system. For example, some 30 to 40 million pastoralists living in the world's driest lands produce animal products in areas that are essentially unsuitable for crop agriculture. Nevertheless, meat consumption can sometimes have unintended negative consequences as well. Land use patterns in agriculture show increased intensity of resource use on arable land. With suitable temperature and rainfall, people can successfully use land to grow crops and produce foods. But where is the arable land located, and how much exists? Certainly not all of Earth's land is used for crops. Of the total land area of Earth, people use about 11% for crops and use another 24% for pastures; 31% is forested. The arable land is heavily concentrated in Canada, the United States, Europe, India, China, and Southeast Asia. The rest of the land surface-about one third-is too cold, too dry, or too steep for plant growth.

The unequal distribution of agricultural resources over the globe is shown clearly if we express the amount of cropland per inhabitant and the available freshwater per inhabitant for each continen. Asia, which already has the most people, has the least agricultural land and freshwater per person. The location of arable land is itself a function of climate, soil type, and type of vegetation that grew in the area before people cleared the land.

Climate, soil type, and vegetation all interacted during the soil formation process and the most productive soils formed in areas of permanent grassland or hardwood forests. Grasslands produce rich soils with abundant organic matter that can remain productive for many decades after farmers convert them to cropland. These soils are quite suitable for annual row crops. The process of converting natural ecosystems into pastures or cropland is now being repeated most dramatically in the tropical rainforests of South America.

However, converting a tropical rainforest into an agricultural field can be very difficult. Most rainforests grow on relatively infertile and highly weathered soil. When people clear such land and burn the vegetation, they expose bare earth to the sun, soil temperatures rise, and the soil organic matter quickly decays, because the vegetation does not replenish it. Because of the high rainfall, the soil erodes easily.

Asia, Latin America, and Africa contain many examples of massive soil degradation caused by clearing away tropical forests and imposing agricultural practices on soils that are totally unsuitable for row cropping. People need new approaches for these soils. Plowing more land may not achieve a greater total of cropland than the 1.5 billion ha presently used.

The FAO estimates that 5 to 7 million ha are being lost each year to land degradation. One problem is that the areas with the greatest population pressures, such as the east coast of China and most of India, have the least land reserves that could be cleared for crops. Other areas-in the former Soviet Union, for example-have good arable land but not enough rainfall. Improved agricultural technologies could greatly expand food production in these regions.

The development of agriculture in different regions of the world has intensified land use. Each stage of intensity has its own characteristics and productivity:

1. *Forest fallow*: Cutting and burning parts of a forest release nutrients contained in the plants; this fertilizes the land, allowing it to support crops. But this fertility lasts only until crops deplete the soil of its nutrients, and the farmers then move on to another part of the forest, repeating the cycle. In the meantime, they leave the original land fallow for up to 20 years, to restore its fertility.

2. *Short fallow*: As human population density grows, grassland replaces the forest and farmers shorten the fallow period to one or two years. This is not long enough to fully restore the soil, to let it support crops, so added fertilizer-often animal manure-is required. Farmers often plant special legumes to accelerate restoration of soil fertility. In addition, they achieve higher yields by manual pest control.

3. *Annual cultivation*: With a still-increasing population, now coupled with market demand, farmers drop the fallow period and grow a crop every year. Now they need constant fertilization and pest control to keep up the yield. Crop rotations increase the sustainability of the system.
4. *Multiple cropping*: This is the most intensive use of the land. Not only do farmers grow several crops on the same land sequentially in one year, but also they grow two or more crops at the same time. Farmers often follow this latter pattern, called intercropping, for the same reason as they rotate crops: One species removes one type of nutrients, and the intercrop removes other nutrients.

Farmers now purchase the following inputs:

– *Farm Machinery*: The process of mechanisation started in the mid-19th century. New farming techniques depended on scientific advances and the industrial production of inputs. Manual labor and animal power were replaced by steam power and later by the internal combustion engine. Farmers used tractors to pull plows or the reaper-binder, which they later replaced with the combine harvester. Milking machines, cotton gins, cotton pickers, sugar beet and potato harvesters, and tomato-harvesting machines all have greatly reduced the need for manual labor. In developing countries small tractors have revolutionised agricultural production and greatly reduced the input of labor.
– *New and Improved Varieties*: Initially, farmers selected suitable varieties of crops for planting the next year. These selected strains were the mainstay of agriculture until the science of genetics emerged in the 20th century. Then it became possible to deliberately

interbreed different strains of a crop to consolidate desirable characteristics in a single strain. The first step is always to evaluate the field performance of known varieties. Scientists have systematically applied plant-breeding principles to improving rice and wheat, and have widely introduced new varieties in Latin America and Asia, displacing local varieties. This process, often called the Green Revolution, has significantly raised crop yields in an era when the amount of land cultivated has remained more or less constant. The introduction of hybrid rice in China and Southeast Asia is raising rice yields even further. About roughly 40% of all increases in crop productivity during the past 50 years stemmed from breeding new varieties. The other 60% improvement has come from managing the crop environment by inputs such as energy, fertilizer, and pesticides.

– *Inorganic Fertilizers*: Originally European farmers relied on manure and crop rotation to maintain fertility. Later they found that ground bones and rock phosphate enhanced crop production, as well as nitrates. The invention of a chemical process that combines nitrogen gas with hydrogen gas to form ammonia, allowed the widespread production of nitrogen fertilizers. This led to a tremendous increase in nitrogen fertilizer use worldwide. Fertilizer applications have been slowly decreasing in developed countries, but in developing countries they are still rising.

– *Herbicides and Pesticides*: Herbicides replaced manual labor for weeding, and farmers used insecticides and fungicides to minimize crop losses. These changes in food production had a great impact on commercial grain farming, on mixed crop and livestock farming, and on intensive wetland rice farming in Asia and

Africa. More complex approaches to weed control and pest control are now replacing heavy use of chemicals in agriculture.

- *Irrigation Technology*: Irrigation has become a very important agricultural input in the past 40 years and accounts for much success in raising food production. Indeed, although only 18% of the world's arable land is irrigated, this land produces 40% of our food. Irrigated land is highly productive. However, this trend cannot be maintained: There simply is not enough water in many areas. Experts now see that water is the resource that will be most limiting for food production in the 21st century.
- *Information Technology*: New innovations in agriculture that use information technology are often called precision agriculture. To obtain maximal yields, farmers of very large farms use remote sensing of their land and their crops by satellite or airplanes to adjust irrigation water, fertilizer application, and genetic varieties. Large tractors equipped with computers can now control row spacing and crop planting densities.

Biotechnological Advances and Agricultural Development

Scientists often breed improved varieties to fit other production-enhancing inputs. For example, after scientists introduced hybrid maize in the United States in the 1930s and nitrogen fertilizers became widely available, agronomists bred maize to respond to nitrogen fertilizers. After the invention of the automatic tomato harvester, tomatoes were bred to withstand the rougher handling to which such machines subject these fruit. These technological advances have led to a style of agriculture that depends on purchased inputs. A survey of U.S. production methods for maize from 1910 to the present clearly shows this dependence.

The technological advances resulted from government sponsored and industrial research, the main thrust of which has always been to develop technological inputs that let farmers minimise per unit production costs. Many of these advances diminished work opportunities on the farm in favor of jobs in towns and cities where the inputs are manufactured. The benefits of the lower production costs (cheaper food) accrued to the consumers and to agribusiness, whereas society as a whole had to bear the penalties (pollution and land degradation). Applying science and technology to crop production in developing countries between 1955 and 1985 resulted in the "Green Revolution." Worldwide food production has been rising by 2.3% annually as a result of the development of high-input agriculture, which in turn results from applying agricultural research and technology. This use of scientific knowledge started about 150 years ago in the developed countries and has resulted in yields taking off spectacularly, as can be charted quite accurately for individual crops. Japan is a classic case of a country turning to high-yield agriculture. By 1900, this island nation was growing crops, mostly rice, on all its arable lands. But the population was still growing.

The government was reluctant to become dependent on foreign food imports lest a hostile nation cut off food from Japan during an international crisis. Short of lowering the people's nutritional standard, the only thing to do was to increase the yield of the rice crop. The government mobilised the country's political, social, and scientific resources, and as a result the yield per hectare tripled between 1900 and 1965. One characteristic of the new crop varieties that raised Japanese rice yields and U.S. wheat yields was that they were short-stemmed and responsive to nitrogen fertilizer. But as with most varieties, scientists had selected and then bred them specifically to grow in the soils and climate of developed regions.

A variety that thrived in the U.S. Northwest had the genetic makeup to do well there, but not necessarily in the drier, hotter regions of the less developed world. There, crop production still relied on local varieties. The genetically improved varieties of wheat and rice that drove the Green Revolution resulted from a targeted crop improvement program at two CGIAR institutes: the International Rice Research Institute (IRRI) in the Philippines and the International Center for the Improvement of Maize and Wheat (CIMMYT) near Mexico City. Farmers adopted the new varieties over a period of about 10 years, repeating a phenomenon that had occurred in the United States some 30 years earlier, when farmers adopted hybrid maize.

In the case of high-yielding varieties of wheat and rice, the farmers had to adopt not only the new seeds, but an entire technology package, including fertilizers, insecticides, weed killers, equipment for irrigation, and tractors to till the land. Indeed, the new varieties let farmers grow two or even three crops per year, potentially increasing the demand for labor. Thus labor-saving technologies (tractors and herbicides) had to accompany the new strains. Furthermore, farmers needed nitrogen fertilizer so the new varieties would yield up to their potential (without fertilizer, they produced the same yields or less, than the landraces they replaced). Adopting these new varieties of wheat doubled and tripled production in Mexico, India, and other countries in Asia. Rice production increased similarly. But the yield increases on farms were never as great as those at research stations. Scientists at IRRI showed that the new varieties could outproduce the old ones by a factor of 3 or 4, yet most farmers in Asia realized increases of only 1.5. This discrepancy in production is not surprising and also occurs in developed countries, where farms seldom match yields obtained in experiment stations. Surveys of farmers who were using the new technology showed that the constraints on higher productivity were poor water control, inadequate nitrogen fertilizers, and diseases. Put another way, the major constraint was inadequate capital to purchase the input package needed to obtain the high productivity of which the seeds are capable.

The fundamental fact of the Green Revolution is that the farmers must manipulate the environment to get the most out of the plants, and too often they have not been able to afford it. Nevertheless, the Green Revolution allowed many developing countries to raise food production and even become food exporters. International agricultural experts saw breeding the high-yielding varieties and applying the industrial inputs as the quickest way to increase production. As a result, agriculture in developing countries came to resemble agriculture in developed countries: more heavily dependent on purchased outside inputs, more capital intensive, and less dependent on labor.

Merely raising production is not enough to eliminate hunger. Because the need for farm laborers decreases as agriculture becomes more technological, the displaced laborers must find other employment. When farmers first apply a new technology-for example, fertilizer or pest control-yield immediately increases as the amount of input rises. A little bit is good, and more seems better! During this phase, the money spent on the technology results in a substantial return. Once the plant's inherent capacity to respond to the input is reached, however, the yield increase slows drastically. Adding fertilizer and other inputs follows the law of diminishing returns: less output for a given amount of input. In the United States, farmers now apply 90-100 kg of fertilizer per ha of cropland, and it is not easy to raise the yield further by adding more.

This figure is 65 kg/Ha for India, so there is still hope for improvement, but fertilizer use is only 4 kg/Ha in Ethiopia and 12 kg/Ha for Costa Rica. Gains in crop productivity should be possible in these countries if researchers can find the proper cropping systems and crop varieties that can take advantage of additional fertilizer.

When all technologies are maximal, agriculture may reach a yield plateau. Some crop physiologists argue that people will always be able to find new crop varieties to escape the yield plateau and take advantage of new and existing technologies. Biotechnology has the potential to allow continual crop production increases. For example, scientists at IRRI and elsewhere are now producing hybrid rice varieties to raise rice yields even further. The CGIAR institutes are catalysts for agricultural research in developing countries. The success of the Green Revolution, and the perceived need for more research on tropical and subtropical food crops, led to the establishment of other research institutes in different countries of the developing World. These institutes have a triple mandate:

(1) to improve the crops assigned to them,

(2) to study farming systems for these crops, and

(3) to create gene banks in which the landraces of the crops under study can be preserved.

Initially the institutes focused on the approach that was successful with highyielding wheat and rice varieties in the 1960s and 1970s. That is, they concentrated on monoculture and on improving crop production by a package of improved varieties and associated technologies to modify the environment. More recently, this focus has begun to change, showing a greater concern for integrating agriculture more with the natural environment. For example, scientists at ICRISAT in India have identified strains of groundnut that produce more than 1,000 kg grain per hectare, even when stressed by drought, compared to the 500-800 kg per hectare the normal varieties produce. Similarly, they have found strains of sorghum that produce two to three times more grain than the present commercial types. One of their goals is to select drought-resistant crop plants to use not only in Asia, but also in the Sahel region of Africa. Scientists at CIAT in Colombia have identified genes in wild varieties of beans that let them resist attack by bruchid beetles. These beetles do enormous damage to the seeds after harvest because they multiply in the dry, stored beans. By crossing a wild bean variety with a cultivated variety, CIAT scientists were able to introduce the bruchid-resistance genes into the cultivated variety.

When the CGIAR system of research institutes was first established, people saw biological and technological research problems as separable from political and social issues. The research institutes generated the necessary innovations and offered these as packages to various national and regional extension services or development authorities. Finetuning these packages to local conditions, and solving social and political problems that resulted from implementing new technologies, were the responsibilities of the implementing agencies or the governments of the countries in question. In this system, people saw agricultural development as an activity that came from the top (the CGIAR institutes) and trickled down to the bottom (the farmers). This approach has worked reasonably well in some countries or areas, and poorly in others. Agricultural development has not often been a high priority for many governments of developing nations, except for developing cash crops that earn foreign exchange. In many African countries, national agricultural research efforts have been weak, in part because governments mistakenly assumed the CGIAR institutes would solve all their agricultural problems, and also because these high-profile institutes attracted all the international funds and the best scientists.

The Green Revolution has done more than raise overall food production. It has also caused the rapid disappearance of the landraces and of indigenous farming systems that may have had much to offer to agricultural researchers. The top-down approach of development implies that agricultural researchers can learn little of value from subsistence farmers. For example, although the institutes have often stressed monoculture, many farmers have been growing more than one crop on the land (either by crop rotation or by intercropping) for centuries. These practices led to a more sustainable agricultural system than did monoculture. Scientists at the CGIAR institutes know they must do more than create new strains to hand over to extension agents. They need to become involved in understanding the tropical agroecosystems at the farm level and to evaluate with the farmers the local varieties (genotypes) and breed varieties for different environments.

Biotechnology is an important tool, but only one of many, to raise crop productivity in the less developed countries. Biotechnology will contribute to the continued rise of crop yields in the 21st century. Between 1960 and 1985 the yield of the major cereals (wheat and rice) rose dramatically in many developing countries, resulting in a 25% rise in the per capita food availability. However, in more recent years this trend has leveled off. But two relatively new factors have arisen to challenge crop producers. First, increased demand for feed grains has followed increased demand for meat. Second, people realise that high-input agriculture can degrade the environment, so that plants must be even better adapted to the natural system. Although people still need ever better genetically improved varieties in combination with new technologies, they also need to more thoroughly understand the sustainability of crop production. Researchers need to devise different strategies for the highly productive regions that are already intensely cropped and where yields have risen dramatically and for the more marginal soils where, as in sub-Saharan Africa, productivity has stagnated and per capita food availability has declined.

Classical plant breeding has been a very important factor in the past successes. A new set of tools has become available in the past 20 years, that-combined with plant breeding-will allow people to produce the genetically improved varieties of the future. This set of tools, which comes under the general title of biotechnology, encompasses a variety of laboratory methods that include:

- Cell, tissue, and embryo culture.
- Clonal propagation of disease-free plants
- Identification of chromosome regions that carry important multigenic traits
- Gene identification and isolation
- Genetic engineering for agronomic traits such as pest and disease resistance or better adaptation to environmental stresses
- Genetic engineering for greater nutritive value
- Genetic engineering to reduce postharvest losses
- Genetically engineered male sterility to facilitate hybrid seed production.

The horticulture industry routinely uses meristem culture and plant regeneration from small parts of the meristem to propagate virus-free planting materials. The discoveries that the genetic material (DNA) of all organisms basically has the same structure and that genes from one organism can function normally after transfer to another organism have opened up the field of genetic engineering. In plants, gene transfer is made possible by a natural process discovered in a soil-dwelling plant pathogen that transfers a few of its genes to plant cells. Molecular plant biologists have manipulated this process so that the pathogen will transfer one or more genes of the scientist's choosing. This technique opens up unlimited possibilities for modifying crop traits. Plant breeders were previously limited to using the genes of the crop's closest relatives, but with genetic engineering they can now use any gene from any organism. It is important to emphasise repeatedly that the genetic engineering and biotechnology can make important contributions but are not the silver bullet that will solve all food production problems. It is but one of the many technologies that people need.

Genetically engineered plants that resist certain insect pests or can tolerate herbicides so that farmers can destroy weeds more efficiently, are being grown in more than half a dozen countries already (Argentina, Australia, Canada, China, India, Mexico, and the United States). The rapid adoption of these new varieties has pleased the biotechnology companies that produce them and the farmers who adopt them because it lowers their production costs, but has aroused some anxiety in the general public.

Some countries in Western Europe and a few localities in the United States have banned genetically engineered (or manipulated, GM) crops. The four major GM crops being grown are rapeseed (canola), cotton, maize, and soybean. Molecular techniques will also greatly facilitate chromosome mapping of important agronomic traits. This will make it easier to follow these traits in the progeny of crosses when the traits have no easily discernible phenotype (characteristic) in the field. It will also greatly speed up classical breeding, which is a rather slow and cumbersome process. Gene replacement (gene therapy) is not yet a reality, but is on the drawing board. It eventually will be possible to take a small chromosomal region that encompasses a whole set of genes, and simply replace it with another region from a plant that is more distantly related.

The effects of intensive agriculture on the ecosystems are causing concern about its sustainability. Growing populations and increasing affluence demand ever-increased output from agricultural systems. This is true whether we are talking about the high-input agriculture of the United States and Western Europe, about nomadic herds in Africa, or about intensive rice cultivation in Asia. Concerns are now being raised about the sustainability of this ever intensifying productivity.

The World Commission on Environment and Development defined sustainable development as "development which meets the needs of the present without compromising the ability of future generations to meet their own needs." To understand sustainability, we must first look at the rate at which people are losing productive soils as a result of present agricultural practices. The FAO of the United Nations has estimated that salinisation, soil erosion, and desertification have degraded a quarter of the world's arable land. Second, there may not be enough energy resources to maintain high-input agriculture. Third, the increasing use of pesticides is arousing concern that people are polluting ecosystems with synthetic chemicals. Fourth, the trend toward genetically uniform crops increases the potential for serious disasters by eliminating the many different strains of a given crop that farmers previously used. Fifth, government policies perpetuate conventional agriculture and discourage farming practices that could make agriculture more sustainable.

The term land degradation denotes the physical and chemical changes that reduce long-term soil productivity. Such changes are sometimes very obvious but are often difficult to measure because they occur over long periods of time. Other improvements, such as more fertilizer or new genetic strains, often compensated for them, so the effects of land degradation do not always show up as decreased yield-at least, not in the short run.Physical degradation takes the form of erosion, the carrying away of soil particles by wind and water. In an annual cropping system, the soil is alternately covered with plants, and then almost totally bare. When the soil is bare, it is exposed to higher wind velocities and to the force of raindrops, which destroy the structure of the uppermost layer of topsoil. The net result is muddy runoff and/or dust storms.

Chemical degradation can take several forms, including

(1) acidification from acid rain,

(2) alkalinisation and salinisation (buildup of salts) as a result of irrigation or the intrusion of saline groundwater into topsoil,

(3) exhaustion of mineral nutrients when farmers don't use enough fertilizer to replace minerals removed with the crop, and

(4) leaching of excess mineral fertilizers into streams, lakes, and groundwater. Pesticides and Herbicides.

Agriculture uses pesticides and herbicides extensively in the developed countries, and their use in developing countries has also increased rapidly. Many problems accompany their use, including the emergence of pesticide-resistant pests, adverse health effects on farm workers, pesticide residues on crops, and pollution of lakes, streams, and groundwater with pesticides and herbicides. Furthermore, the realisation that pesticides have not diminished the proportion of the crop lost to pests is prompting many people to try alternate pest control methods. There is renewed emphasis on pest-resistant varieties and biological pest control.

Loss of Biodiversity

A main feature of the high-input system of agriculture is genetically uniform crops. These reliably produce high yields if used with appropriate inputs. But their use has several drawbacks. First, when farmers start planting seeds furnished by seed companies or state agencies, they stop using and usually discard their own varieties, which were well adapted to the microenvironment on their farm. Local varieties contain valuable genes that are thus lost to plant breeders. Second, genetic uniformity means crops are more susceptible to sudden outbreaks of disease. When a pathogen mutates and overcomes the plant's resistance, the entire crop over a wide geographic area can be ravaged.

An excellent example is the potato blight that ravaged the Irish potato crop in the 1840s. Sometimes government policies actively promote high-input agriculture and discourage sustainable practices and technologies. Such policies are difficult to change because large and powerful lobbies, including farm organisations and industry advocates, have vested interests in such policies. To protect their resource base-the soil-and to decrease costs, farmers have begun to adopt alternative practices. Taken together, these practices constitute sustainable agriculture, which differs from conventional agriculture not so much by the practices it rejects, but by the practices it incorporates into the farming system. The more widely used term organic farming describes two major aspects of alternative agriculture:

(1) substituting manures and other organic matter for inorganic fertilizers and

(2) using biological pest control instead of chemical pest control.

The objective of sustainable agriculture is to sustain and enhance, rather than to reduce and simplify the biological interactions on which production agriculture depends. Alternative agriculture is not a single system of farm practices, but encompasses many farming systems variously called organic, biological, lowinput, regenerative, or sustainable systems. Such systems emphasise management practices as well as biological relationships between organisms; in addition, they take advantage of naturally occurring processes such as nitrogen fixation.

Alternative or sustainable agriculture is any food or fiber production system that systematically pursues the following goals:

- More thorough use of natural processes such as nutrient cycles, nitrogen fixation, and pest-predator relationships in agricultural production
- Reduced use of off-farm inputs with the greatest potential to harm the environment or the health of farmers and consumers
- Greater productive use of the biological and genetic potentials of plant and animal species
- Improved match between cropping patterns and the productive potential and physical limitations of agricultural lands, to ensure long-term sustainability of current production levels
- Profitable and efficient production, with emphasis on improving farm management and conserving soil, water, energy, and biochemical resources

— Studies of alternative agriculture show that the farms derive sustained economic benefits and are not necessarily at a disadvantage. Although yields are often lower, costs are also considerably lower.

Climate Change

The weather is the one factor with which people are familiar that profoundly affects the growth of plants. Everyone knows that to grow plants need sunlight and rain and that both are quite variable around the world. Weather-defined as short-term (less than a few weeks) variations in the atmosphere-affects the year-to-year yield of crops. The longer-term variations that comprise climate determine the geographic conditions where crops can grow at all. Short-term variations caused by the weather that affect crop yields occur even in the modern era of technology-based agriculture:

1. Moisture stress is caused by insufficient, too much, or ill-timed rainfall. Plants often have precise water requirements, in terms of both amount and time. If these requirements are not met, yields and/or quality suffer.

2. Temperature stress occurs commonly when the temperature is either too low or too high for optimal growth. A short period of stress can severely depress growth later. Sudden cold snaps injure plants because they do not have enough time to become acclimatised to the cold weather. Moisture stress often accompanies high-temperature stress.

3. Natural disasters occur, such as cyclones, hurricanes, and hailstorms. Although these are more common in such Asian countries as Sri Lanka and Bangladesh, they can also cause severe damage in developed countries by physically harming crops and the soil that supports them.

Yields of many crops vary considerably year to year in many locations. The probability of a staple crop yielding the same year after year can be quite low. Over much of North Africa and the Middle East, for example, yields fall significantly below expectations in half of the crop years. A similar situation exists in the countries of the former Soviet Union. Complex statistical models, proposed to explain crop yield variability in other situations, show that in many cases climate is the major factor. When things go well (the yield trend is a smooth upward or constant curve), it is because the weather is unusually benign. Different regions of the world are affected differently by climatic variation. Some of these effects are as follows:

1. *Temperate regions*: The north and south temperate regions produce 75% of the world's wheat and maize. The most prominent climatic factors affecting yield are moisture stress and temperature extremes. Wheat and soybeans are more drought resistant than maize.

2. *Humid tropics*: Ranging from fertile valleys to jungles to floodplains, these regions are heterogeneous and show great variability of year-to-year yields. The major determining factor here is rainfall. There are pronounced wet and dry seasons, with the wet season often being in the form of a monsoon, with very heavy, sustained rainfall. But the unpredictability of the intensity of the monsoon, combined with poor water-holding capacity of soils and high rate of evaporation, lead to unstable water supplies for agriculture. In addition, torrential rains on lowland areas, such as river deltas, inundate the crops.

3. *Semiarid tropics*: These regions (for example, the Sahel south of the Sahara, in Africa, and much of India) have a long history of periodic crop failures and resulting famines. The major problem here is rainfall

> or, more precisely, the lack of it. Most of the rain falls in a two- to five-month period (April-October in the Northern Hemisphere, October-April in the Southern Hemisphere), when temperatures are at their yearly peak. This leads to extensive evaporation and less water availability. The sequence of events in growing a crop here is exquisitely sensitive to the annual rainfall cycle. Farmers time the growth of the crop to coincide with the maximum available water. If this period is very short, so is the growing season. In these areas farmers often plant two crops: one that will yield some food if the rains fail to come, and one that will yield abundantly if they do come.

This relationship of agricultural practices to climate is an ancient one. Every culture has its deities and stories relating the cycle of the seasons to food production. Human activities can substantially alter the global environment and thereby negatively impact our ability to grow crops. The public is now keenly aware of the effects of pollution on society, and generally considers industry to be the main source of pollutants. However, in the past 30 years agriculture has also become an important environmental polluter. The book Silent Spring, written by biologist and science writer Rachel Carson in 1962, first awakened public opinion to unintended effects that pesticides can have on the environment. Although organochlorine insecticides then used have since been phased out, they did considerable damage both in developed and developing countries. These pesticides prevented crop damage by insects, but in Asian rice paddies pesticides kill fish, shrimp, and crabs, important sources of protein for poor people.

Burning vegetation to clear forestland or to encourage grass growth in savannas is an important source of global air pollution. When vegetation burns, some nutrients (potassium, phosphate, and calcium) return to the soil, but nitrogen, sulfur, and carbon disappear into the atmosphere as gases. In addition, burning releases particulates into the air. The sulfur dioxide and nitrogen oxides, in combination with those same two gases released from the burning of fossil fuels, help acidify soil via acid rain. In Europe and North America acid rain is killing some forests, and in China it has shrunk rice and wheat harvests, with most of the damage occurring at the beginning of the rainy season.

Burning biomass (vegetation) annually releases about 1 to 2 billion tons of carbon into the atmosphere as carbon dioxide. This amount adds to the 2 billion tons of carbon released by burning fossil fuels. Carbon dioxide is a greenhouse gas, which means it contributes to warming of Earth. Anaerobic decay of organic matter in marshes produces methane, another greenhouse gas. Rice paddies are an important source of methane worldwide, and increased rice cultivation in Southeast Asia has much increased methane production. These gases are responsible for the greenhouse effect, which warms the surface of Earth. If greenhouse gases were completely absent from the atmosphere, Earth's surface would be a chilly -18°C. Their presence maintains Earth at a higher temperature, making human life possible.

The temperature of the planet has been rising slowly-about 0.3 to 0.6°C during the past 100 years, and is expected to rise another 0.4 degrees C by 2020. A2°-C rise by the end of the 21st century is possible. Although this slow global warming could be a natural phenomenon, most experts believe it is the direct consequence of the rise in emissions of greenhouse gases. A likely result of global warming is greater variability in the weather and more extreme weather conditions.

Rainfall may be more variable, making crop production even more variable than it is now. Other gases, especially nitrous oxides and the now banned chlorofluorocarbons formerly used as refrigerator coolant, destroy the protective ozone layer in the upper atmosphere. Ozone shields Earth from destructive ultraviolet solar radiation. Closer to Earth, nitrous oxides emitted from cars or released by burning vegetation help create photochemical smog, of which ozone is a significant component. Even at quite low concentrations, ozone inhibits plant growth and recent experiments suggest that the ozone levels around cities and in some rural areas are now high enough to impair crop yields. Scientists may need to take this into account when breeding new crop varieties.

Sustainable Use of GM Crops

When potential risks to the environment from new genetically modified (GM) crops are discussed, the focus is generally on risks of genes escaping to wild relatives that become "superweeds," risks of crops themselves becoming weeds, and the potential that toxins produced by the GM crops will harm nontarget organisms . These are all tangible risks, that can be diminished if taken seriously. Their worst-case negative impacts on the overall environment in developing countries are relatively small compared to the impact on the environment of rural populations without food security and living in poverty. Because poverty can lead to environmental degradation, a potential chain of interactions between genetic modification, increased yield, variation in yield, food security, and environmental degradation. The new genetic modification has great potential for increasing yield and decreasing yield variation, but that such accomplishments will require vigilance by scientists and society.

Giving farmers the right kind of seeds can never ensure food security, but giving them the wrong kind of seeds always can make things worse. Social scientists know that you cannot alleviate poverty just by growing more food, but agricultural scientists are always being pushed by philanthropic and development organisations to produce more food because food production is a necessary but not sufficient condition for food security. This has led to increased yield becoming a focus of international attention. About 50 years ago agricultural scientists were told that the growing world population would demand that substantially increase crop yields.

Amazingly, agricultural scientists, with the financial aid of many government and private organisations, did a reasonable job of meeting this demand. Today agricultural scientists are being told to do this again because world population growth by the year 2025 or so will demand that increase production by 30-50 percent. One thing that crop and animal breeders have learned over the years is that when you select strongly for improvement in one trait of an organism, be it speed in a race horse, appearance in a dog, or yield in a crop, there are tradeoffs. The race horse may be frail, the dog less intelligent, and the crop less capable of dealing with stress. There has always been support given to programs that breed for stress resistance, but these programs are often funded with the goal of increasing yield rather than decreasing variation in yield.

A number of recent reviews have attempted to determine if the variation in crop yields have changed during the 20th century. The conclusions of these assessments indicate that on a global and continental level overall adjusted variation in yield has not generally increased, but that the component associated with genetic crop improvement has increased. This means that the variation in yield is being pushed upward by the varieties of crops grown, but that other factors such as improved irrigation may be balancing out this increase caused by breeding. Because at the local level, drought, flooding, and pests vary significantly from year to year, cultivars bred with resistance or tolerance to any of these disruptive factors would decrease yield variation and could also result in increased average yield. Genetic engineering has improved and should continue to improve, these traits, but care must be taken in how this is done if a major goal is decreased variability in yield.

Pest Resistance Management

Crop breeders have long known that some conventionally bred cultivars with resistance to insect and microbial pests may perform wonderfully for the first few years after they are deployed commercially, but then fail miserably in controlling the targeted pests in later years because the pest has evolved a way to cope with the resistance mechanism in the cultivar. Sometimes there is a slow decline in effectiveness of the cultivar, but in other cases the onset of control failure is rapid and unpredictable. If you are a subsistence farmer, the failed performance of such cultivars can mean hardship, especially if the cultivar had previously performed well and long enough for you to gain confidence in it.

Indeed some of the criticisms of the Green Revolution of the 1960s and 1970s centered on rice cultivars that were rapidly adapted to by insect and microbial pests. For example, brown planthopper populations adapted to the single-gene resistance in the first Green Revolution rices within 2-3 years of their widespread cultivation, and single-gene resistance to the rice blast fungus has been notoriously unstable. The longevity of cultivars with single blast resistance genes in Japan has been less than 3 years. In industrial countries, breeders and seed producers sometimes try to deal with pest adaptation to widely used crops that have one resistance mechanism by maintaining, in reserve, replacement cultivars with different resistance mechanisms, for example wheat rust. These systems are sometimes able to replace cultivars in a single season as was the case with the southern corn blight epidemic in the USA. In developing countries, instituting such a system for subsistence crops is difficult or impossible because of limited infrastructure and resources.

Not all pest-resistant cultivars are rapidly adapted to by their target pests. Entomologists and plant pathologists have worked hard to predict whether a specific resistant cultivar is likely to work well for a long time under field conditions. This characteristic called "durable resistance" has proven to be partially predictable, but many plant pathologists are only willing to judge the durability of a specific type of pest resistance in retrospect.The general problem of pests adapting to any approach used to control them has been the bane of agriculturalists for centuries. Weeds, pathogens, and insects have all overcome various cultural, chemical, and biological approaches used for their control.

Over 500 insect species are known to have adapted to at least one insecticide, and it often takes less than three years for this adaptation to evolve. In many developing countries this can severely disrupt food production because replacement insecticides are often not available, and the beneficial insect populations have been decimated by insecticide use.In 1997 and 1998, there was a tragic series of over 400 suicides among cotton farmers in Andra Pradesh, India in response to crop failures that were in part the result of pest adaptation to insecticides. The farmers were heavily in debt because of several seasons of crop failures, caused by irregular rainfall and heavy infestations of the insect pests Spodoptera litura and Helicoverpa armigera.

Application of large doses of highly toxic insecticides such as monocrotophos and methomyl were not effective because of pest resistance to these compounds, and their toxicity to predatory and parasitic arthropods which otherwise could have provided some level of natural biological control. In the 1970s entomologists, plant pathologists, and weed scientists began a concerted effort to use knowledge of evolutionary biology and population genetics to develop strategies for slowing the rate at which pest populations evolved adaptations to control tactics such as pesticides and pest-resistant crops. This approach called "pest resistance management" now seems highly appropriate for crops developed using genetic engineering, because there is good reason to predict that some approaches to the development and deployment of engineered pest-resistant crops will last much longer than others.

When new genetically improved crops that expressed insecticidal proteins from Bacillus thuringiensis (Bt) were first developed, there was much concern in the United States about insects adapting to these toxins. Unlike conventionally bred resistant crops, where a resistance mechanism can only be moved within a single crop species, the Bt toxins were being moved into multiple crops, so insects that fed on more than one crop would get multiple exposures. Unlike insecticides that are sprayed only during some time periods in the season when pest pressure is high, the newly developed crops produce the toxin all season long, so all insects in a population can be exposed to the toxin. Everything known about pest adaptation indicated that overuse of such crops could give great control for a limited number of years followed by failure.

In the United States there was one other pertinent fact about Bt crops. B. thuringiensis, the bacterium that was the source of the toxin genes in the crops, has long been sprayed on crops by organic farmers and others as an alternative to chemical insecticides. Organic farmer groups and their supporters protested that the overuse of Bt toxins in genetically engineered crops, and the subsequent development of adapted pests, would leave them without an effective pest control tool. This highlighted two issues: one was the plight of the organic farmer and the other was the unique, environmentally benign nature of Bt toxins compared to conventional pesticides. A set of Bt toxins, sometimes referred to as Bt endotoxins, were known from previous uses to be effective at killing either some caterpillars or some beetle species, but they had no effect on almost all other species. From an environmental perspective these are wonderful toxins, and unless other toxins with this high target specificity can be quickly found, the overuse and loss of Bt toxin efficacy in transgenic crops could send cotton and potato farmers back to spraying environmentally disruptive chemicals.

Bt-Resistance Management

Prior to the commercialisation of any genetically improved crops, the US-EPA held meetings of Scientific Advisory Panels to get advice from experts outside EPA regarding risks of genetically improved crops. One of the recommendations of these panels was to institute resistance management programs. When EPA granted conditional registrations for the first Bt corn cultivars in 1996, one of the conditions was the development of resistance management plans by the year 2000. The EPA felt that such plans were not immediately needed because they expected adoption of these Bt cultivars to proceed more slowly than it actually did. The conditional registration for Bt cotton included a resistance management plan, but this plan is now being reexamined because it lacks rigor.

More recent conditional registrations of newer corn cultivars have included more stringent resistance management plans. The imposition of resistance management plans is something new for the US-EPA and the agency has been gaining sophistication in this area over time. In 1998 the EPA convened a Scientific Advisory Panel to reassess the issue of Bt resistance management. The report of this panel laid out some clear recommendations to the EPA. After considering a number of potential resistance management strategies, the panel recommended that "a refuge/high dose strategy must be employed for target pests within the current understanding of the technology." They added that "regulatory strategies should serve to provide growers with a sustainable approach that encourages compliance." These were important recommendations worthy of careful examination.

The high dose portion of this strategy is most easily understood by analogy to the use of antibiotics. When doctors prescribe antibiotics they often give their patients the admonition that even if they feel completely cured after three days, they should continue to take the antibiotic for the full time prescribed. The reason for this is to produce a high dose of antibiotic for an extended time period that will kill even those rare bacterial cells that have a mutant gene conferring partial tolerance of the antibiotic. After three days you may have killed 99 percent of the targeted bacteria, but if the 1 percent that survive have a gene that confers partial tolerance and are transmitted to another individual, his or her infection will be more difficult to treat.

More importantly, when that next individual takes the antibiotic, the partially tolerant bacteria may evolve even higher tolerance if among the millions of bacteria involved in the infection there are a few bacteria with other mutations that add to the tolerance conferred by the initial mutation. When a patient takes an antibiotic for the full period prescribed, the expectation is that even the partially tolerant bacteria will be killed. As long as it takes more than one evolutionary step to result in complete tolerance of the antibiotic, the prolonged use of the antibiotic should derail the adaptive process by inhibiting the first step. The use of a high dose of Bt toxin in crops serves a similar purpose. In all cases studied to date it takes more than one gene, or at least more than a single copy of a gene to confer high tolerance of Bt toxins. When Bt crops are first commercialised it is estimated that about 1 in 1000 individuals may carry one copy of a gene for tolerance of Bt, and only 1 in 1 million would carry the two copies needed to achieve a high level of tolerance. The high dose approach is set up to ensure that each plant that produces Bt toxin produces enough to kill most of the partially tolerant individuals.

But if the high dose is used by itself, some insects out of the billions that can infest a local area may have two copies of the gene. If they survive and mate, the Bt crops could rapidly lose effectiveness. This is where the refuge part of the "refuge/ high dose" approach comes in. All of the current target insects for Bt crops are obligately sexual. That means that they must mate to reproduce, and that their offspring obtain half of their genes from each parent. If a small portion of a farm is planted to a cultivar that does not produce Bt toxin this area serves as a refuge for Bt-susceptible insects. Because the highly tolerant insects are expected to be so rare, they are likely to mate with susceptible insects produced in the refuge. The offspring of these matings will have only one gene for tolerance, and so will be killed if they feed on a Bt-producing plant. By combining the refuge and the high dose, this strategy derails the evolutionary process as long as more than one gene copy is required to survive the high dose. Pests can eventually adapt to such a strategy but the time period required can be 10 to 100 times longer than expected if this strategy is not implemented.

The 1998 EPA Scientific Advisory Panel was clear about what constitutes a high dose and what constitutes an effective refuge. They defined a high dose as 25 times the amount of toxin needed to kill susceptible target insects. They concluded that an effective refuge existed when for every insect with a resistance gene produced in the Bt crop there would be 500 susceptible insects produced that could mate with the resistant insects. These are stringent requirements and they work in concert. If a crop does not quite produce a high dose, the expected number of insects with at least one resistance gene increases. This results in the need for a larger refuge to produce the 1:500 ratio. How do these recommendations line up with Bt crops that are now on the market in the United States? With Bt potato the data indicate that there is a high dose for the target pest, Colorado potato beetle. However, it is not confirmed that farmers are planting effective refuges. With corn, most cultivars produce a high dose for the European corn borer, but not for the corn earworm.

Refuges currently appear large enough for the European corn borer, in part because of lack of full adoption of Bt corn, but the refuges may sometimes be too far from the Bt corn to allow insects from the refuge to cross-mate with insects from the Bt crops. In cotton there is a high dose for the tobacco budworm but not for the corn earworm . Proposed refuges of about 10 percent are expected to be sufficient for the tobacco budworm, but may not be sufficient for the cotton bollworm. There appears to be a high dose in cotton for the pink bollworm, but this has not been completely confirmed. There is certainly room for improvements when the producers of Bt crops present their new resistance management plans to the US-EPA in 2000.

At a recent USDA/ EPA workshop, there was debate as to whether similar resistance management strategies would be required in China. The expert statement was conditioned on the assumption that no Bt corn was grown in China. If that assumption held then the targeted Chinese pest on Bt cotton, Helicoverpa armigera, could utilise non-Bt corn to produce the Bt-susceptible insects. It appears that there now will be Bt corn grown in China, and the Bt cotton that is being planted in China does not have a high dose for the target insect. This is not the kind of scenario that is likely to retard the evolution of adapted pests. If Chinese farmers and administrators come to rely on Bt corn and cotton in the next few years using the current technologies and risk management, there is a clear risk that yields will show variation over time. There is economic pressure in many developing countries to adopt Bt crops that were developed to control U.S. pests. If these crops are sold as "second hand" cultivars to these countries, it is hard to imagine that they will usually be effective at thwarting pest adaptation.

Unless there are careful contingency plans to deal with the eventual failure of these cultivars, their adoption by developing countries could lead to higher than necessary yield variation. For example, it has been argued that situations such as that in Andra Pradesh demonstrate the urgent need for the release of Bt cotton in India. Two kinds of Bt cotton have been field-tested in India: one from the Monsanto Company and one produced by an Indian research institute.Many farmers will be eager to adopt Bt cotton, which can help to control S. litura and H. armigera. The availability of Bt cotton may also attract many new farmers to invest in growing cotton. However, if a carefully designed resistance management plan is not implemented, the farmers may suffer another severe setback in a few years should the pests adapt to the Bt cotton. In desperation, these farmers may again turn to highly toxic insecticides, repeating a tragic cycle.

By the time insect pests adapt to Bt crops, biotechnology will develop replacements. If it was easy to develop replacements, competitive market pressures in the U.S. cotton and corn seed trade would have already resulted in alternative toxins being produced. The only even marginally novel toxin in corn is the Cry9C Bt toxin that AgrEvo has marketed, but because of health concerns, the US-EPA has approved its use only in corn grown for livestock feed. It has been said that the people with the most to lose from pest adaptation to Bt toxins are those who own the companies that own the genetically engineered seeds, but the aggressive stances of some companies against implementation of resistance management plans recommended by US crop scientists does not, however, indicate sincere concern. An explanation for this discrepancy could be that as with the pesticide market, most of the important profits from Bt crops are expected to come in the first years of widespread use, and resistance management could interfere with these early profits. Public and commercial interests are therefore expected to classh.

Bibliography

Adelberg, Edward A., (1960) (ed.) *Papers on Bacterial Genetics.* Boston: Little, Brown and Company.

Ausubel, F.M. et al., (1989) *Current Protocols in Molecular Biology.* New York: John Wiley and Sons.

Ausubel, F.M. et al., (1989) *Short Protocols in Molecular Biology.* New York: John Wiley and Sons.

Blatt, Robin J.R., (1993) "The Human Genome Project" in association with MCET The Mass LearnPike, Cambridge, Massachusetts.

Broach, J.R., (1981) *The Molecular Biology of the Yeast.* Cold Spring Harbor: Cold Spring Harbor Laboratory.

Brock, T. and Madigan, M, (1992) "Gene manipulation and genetic engineering," *In: Biology of Microorganisms 6th Edition.* Englewood Cliffs, NJ: Prentice Hall.

Bud, R. (1989) "Janus-faced biotechnology - An historical perspective." *Trends in Biotechnology.*

Clark, J.M. and Switzer, R.L., (1977) *Experimental Biochemistry.* New York: W.H. Freeman and Company.

Curtis, H. and Barnes, N., (1989) *Biology.* New York: Worth Publishers.

Diamond, J., (1993) *The Arrow of Disease.* Discover. October, 1993.

Dixon, Dougal, (1981) *After Man - A Zoology of the Future.* New York: St. Martin's Press.

Dodds, John H., (1985) *Plant Genetic Engineering.* New York: Cambridge University Press.

Friedberg, E.C., (1985) *DNA Repair.* New York: WH Freeman and Company.

Goodman, D.C., (1987) *From Farming to Biotechnology: A Theory of Agro-industrial Development*. Oxford: Blackwell.

Horn, Toby M., (1993) *Working with DNA and Bacteria in Precollege Science Classrooms*. Reston, VA: National Association of Biology Teachers.

Mullis, Kary B., (1990) "The Unusual Origin of the Polymerase Chain Reaction." Scientific American. April: 56-61.

Old, R.W. and Primrose, S.B., (1989) *Principles of Gene Manipulation 4th Edition*. London: Blackwell Scientific Publications.

Panicali, D. and Paoletti, E., (1982) "Construction of pox viruses as cloning vectors: Insertion of the thymidine kinase gene from herpes simplex virus into the DNA of infectious vaccinia virus." Proc. Natl. Acad. Sci. 79:4927-4931.

Raven, P. H. and Johnson, G.B. (1992) *Biology*. St. Louis: Mosby Year Book

Retzer, W.J., (1991) Biotechnology Workbook. Englewood Cliffs, NJ: Prentice Hall.

Seabrook, J. (1993) "Tremors in the Hothouse." The New Yorker. July 19:32-41.

Silber, S. (1991) "How to Get Pregnant with the New Technology". Warner Books.

Slater, R.J. et al., (1986) *Experiments in Molecular Biology*. Clifton, NJ: Humana Press.

Torrey, J.G., (1987) "The Development of Plant Biotechnology." American Scientist. 73:354-363.

Towle, Albert, (1988) *Modern Biology*. Austin, TX: Holt, Rinehart and Winston.

Walden, Richard, (1988) *Genetic Transformation in Plants*. Milton Keynes, England: Open University Press.

Index